To Jay Baggett

Warmest Wishes,

Front Cover Photo: Alfred A. Stratton (1845-1874) a blacksmith from Jamestown, New York, joined the 147th New York Infantry on 19 August 1863. Above the elbow, his arms were "carried away by a solid cannon shot" fired from C.S. Army defenders at Petersburg, Virginia, on 18 June 1864. Four months later Pvt. Stratton was promoted to sergeant and medically discharged from the U.S. Army on 3 October 1864.

Stratton married the next year and served as the pastor of Washington Street Episcopal Church in Brooklyn. He posed for this photograph on Christmas Eve in 1869. Amputees usually sold prints of such photographs to supplement their military pensions. At age 29, Stratton died from consumption. He is buried in Mount Olivet Cemetery at Washington, DC. He and other such maimed men are among the many hundreds-of-thousands of casualties of war.
[photo courtesy of U.S. National Museum of Health and Medicine]

– *Current Books by Marion Sturkey* –

Warrior Culture of the U.S. Marines (3rd Edition)

No Greater Love: The Groucho Marx Battle

Bonnie-Sue: A Marine Corps Helicopter Squadron in Vietnam (2nd Edition)

Murphy's Laws of Combat (2nd Edition)

Mayday: Accident Reports and Voice Transcripts from Airline Crash Investigations

Mid-Air: Accident Reports and Voice Transcripts from Military and Airline Mid-Air Collisions

Military Monuments and Memorials in South Carolina

Plum Branch: Heaven in South Carolina

Gone But Not Forgotten (2nd Edition)

Casualties of War

– *Out of Print* –

Gone But Not Forgotten (1st Edition)

Gone But Not Forgotten: An Introduction

Bonnie-Sue: A Marine Corps Helicopter Squadron in Vietnam (1st Edition)

Murphy's Laws of Combat (1st Edition)

Warrior Culture of the U.S. Marines (1st and 2nd Editions)

Casualties of War

Marion F. Sturkey

Heritage Press International

Casualties of War

First Edition

Library of Congress Control Number: 2017916563

ISBN: 978-0-9913011-3-3

Heritage Press International
204 Jefferson Street
P.O. Box 333
Plum Branch, SC 29845
USA

Manufactured in the United States of America

– *Table of Contents* –

Preface. ix
The Incalculable Toll of War. 1
Rise of the Cotton Kingdom. 4
The War for Southern Independence. 14
Wounded and Sick Confederate Soldiers. 42
Prisoners of War. 69
Letters, Diaries, and Memoirs. 93
The Southern Icon. 121
The Thorny Issue of Slavery. 146
Nautical Warfare. 164
Almost Home – The Sultana Disaster. 200
The Soldier Atop the Memorial. 211
The Tragedy of the Reconstruction Era. 224
Casualties of War. 243
Building a New Nation. 258
About the Author. 271
Index. 272

God of our fathers, known of old,
Lord of our far-flung battle line,
Beneath whose awful hand we hold
Dominion over palm and pine –
Lord God of Hosts, be with us yet,
Lest we forget – Lest we forget!
[Rudyard Kipling, Recessional, 1897]

– Preface –

The American Civil War created the greatest human catastrophe in United States history. Many thousands of books have chronicled military aspects of that tragic war.

Without *politically correct* psycho-babble this project focuses upon core issues and their consequences, not upon military events. Primary emphasis is directed toward casualties, societal losses, and the common Confederate soldier.

in memory of

Pvt. Marion Benjamin Sturkey, CSA (1837-1920)

Sgt. Daniel Presley Sturkey, CSA (1840-1864)

Pvt. William Oscar Sturkey, CSA (1842-1921)

Pvt. Thomas Jefferson Sturkey, CSA (1846-1906)

Casualties of War

None can realize the horrors of war, save those actually engaged. The dead lying all about, unburied to the last. My God! My God! What a scourge is war!

[Pvt. Samuel Johnson, CSA, in a letter to his parents, 1864]

The Incalculable Toll of War

Not for fame or reward, not for place or rank,
Not lured by ambition or goaded by necessity;
But in simple obedience to duty as they understood it,
These men suffered all, sacrificed all,
Dared all – and died.
[Randolph McKim, CSA chaplain, later inscribed on the Confederate Soldiers Memorial in Arlington National Cemetery]

From a rural cotton farm in western South Carolina three young brothers answered the call of their country. They enlisted in the new Confederate States Army and marched away to war.

Each brother sacrificed his life of hope and promise on the altar of the Confederate States cause. One by one, each of them died. Via rail and wagon the C.S. Army shipped their bodies back to the family farm. The boys' father, John Briggs, had died in 1859 before the war began, but their mother still lived. One by one she buried her three boys side-by-side in the little community cemetery and erected a tombstone at the head of each grave. Epitaphs inscribed on the three tombstones are displayed below:

Sacred
to the memory of
HENRY BRIGGS
of Company I, 7th Reg*[iment]*, S.C.V*[olunteers]*,
who died at Chimbarazo
Hospital, Richmond, Virginia,
in the 25th year of his age.
He was an exemplary
member of the Baptist Church.
Dearest Henry, thou hast left us,
Here thy loss we deeply feel.
But 'tis God who has bereft us,
He can all our sorrows heal.

Sacred
to the memory of
Corporal
RICHARD P. BRIGGS
who was wounded at the
Battle of Sharpsburg, Mary-
land, *[on the]* 17th of Sept. 1862, and
died the 27th *[of September 1862]* in the 22nd year
of his age.
He was a member of Comp*[any]* B,
Hampton Legion.
The din of battle he will hear
no more.
The war-worn soldier's toils
are o'er.
The dew of his youth to his
country was given.
May he sleep in peace 'till we
all meet in Heaven.

Sacred
to the memory of
THOMAS J. BRIGGS
of Comp*[any]* B, Hampton Legion.
He left home July 6th, 1862,
and died at General Hospital
No. 1 *[on]* Sept. 25, 1862, in the
18th year of his age.
He was the last of three brave
sons who went forth to battle
for his country.
So fades the lovely, blooming flow'r,
Frail, smiling solace of an hour.
So soon our transient comforts fly,
And pleasure only blooms to die.

The three Briggs brothers were gone. The war had snuffed out their lamps of life. Yet, the war exacted a much greater toll.

The boys' mother, Eliza Briggs, struggled until she died in 1896. Friends buried her beside her husband and her three boys. The little farming community grew and prospered, and eventually it became a municipality. The once-small cemetery grew dramatically as well, but never again would anyone named Briggs be buried there. The entire family had been swept away by the war.

The thriving community never again would include a Briggs wife or mother. There never would be more Briggs children, no grandchildren, no brothers or sisters, no nieces, no nephews, no posterity. Everything the family had, or ever would have had, was lost. Family hopes, plans, dreams, aspirations, and potential had been snatched away. Further family contributions to society, to the church, to the state, and to the nation never could materialize.

Thus, the community had lost much more than the three Briggs brothers who were consumed by the war. The total loss is beyond measure. Yet, that incalculable loss is the true toll of war.

These three tombstones mark graves of the three Briggs brothers.

Rise of the Cotton Kingdom

> Without the firing of a gun, without the drawing of a sword, should *[the United States]* make war upon us, we could bring the whole world to our feet. What would happen if no cotton was furnished for three years? . . . England would topple headlong and carry the whole civilized world with her. No, you dare not make war on cotton. No power on earth dares *[to]* make war on cotton. Cotton is King!
> *[Sen. James Hammond, speaking in the U.S. Senate in reply to a senator from New York, 4 March 1858]*

The Cotton Kingdom: As the Nineteenth Century dawned, farming was the primary occupation in the southern United States. Most families labored on small subsistence farms. Exclusive of tobacco there was no viable cash crop. Yet, the new cotton gin developed by Eli Whitney was poised to revolutionize southern farms.

With a hand-cranked cotton gin a farmer could produce 60 times as much cotton lint as a person who had to separate the lint and seeds by hand. Fertile southern soil, warm summer days, and moderate rainfall created the perfect combination for growing cotton. Cotton plants needed nothing else except air to flourish. Even fields too poor for growing corn proved to be excellent for growing cotton.

> The cotton belt spans the southern half of the United States. Its growing season of approximately 150 to 180 days is the longest of any annually planted crop in the country.
> *[editor, Cotton: From Field to Fabrics, 2009]*

The new cotton gin gave southern farmers a lucrative cash crop, and almost immediately they devoted all available land to cotton production. A decade earlier in 1791 the entire United States had produced less than 2 million pounds of raw cotton. Yet, the cotton gin enabled cotton production to soar to 48 million pounds by 1810. By 1860 fifty years later, production would tally an astounding 1,650 million pounds. Almost all of this cotton would come from southern states, and they became known as the Cotton Kingdom.

Farmers on small tracts used family members to pick their own cotton. Then they used a hand-cranked gin to produce salable cotton fibers. Cotton production made them wealthy and provided both necessities and luxuries for their families. The majority of small farmers never owned slaves. Big plantations required more manual labor, so large plantation owners used slaves to work in the fields.

Rise of the Cotton Kingdom ushered in dramatic social, cultural, and economic changes throughout the South. Vast sums of money flowed into the hands of cotton growers. Production spread across southern states: Texas, Mississippi, Louisiana, Alabama, Georgia, South Carolina, North Carolina, Arkansas, and Virginia. Within 50 years these southern states would produce 68 percent of the entire world's raw cotton supply.

Storm Clouds Gather: Potential economic disaster arrived in the form of new federal tariffs and restrictive regulations. A brief look at the history of the U.S. Constitution is necessary.

After the American Revolutionary War the states had needed a new document to replace the old Articles of Confederation. Union of the states was fragile, for northern and southern states held vastly conflicting views. Agricultural southern states had lobbied for baronial rights where yeoman farmers and plantation owners could chart their own course in life. In contrast, industrialized northern states had wanted a strong central government that, while not totally socialist, would impact many aspects of daily life. The Founding Fathers compromised. They adopted a new constitution in 1787, to become effective in 1789, to avoid chaos and anarchy. Yet, neither the North nor the South was fully satisfied with the result.

By the 1820s many northerners resented the growing affluence of Cotton Kingdom states. States in the South lobbied for free trade, but the federal government responded with ever-increasing tariffs. The despised Tariff of 1828 established a 50 percent levy on many items, primarily finished cotton products from England. American textile plants in the North were not as sophisticated as those in England. Without the tariff, imported English cotton products were far less expensive. The tariff was designed to protect the domestic textile industry in the North from low-cost British imports.

In response to the tariff Britain reduced its imports of raw cotton from the South. Southern farmers watched in dismay as the price of cotton dropped. The South was outnumbered in the U.S. House of Representatives by the densely populated North (by 1860 the free population of the North would swell to 22 million; the free population of the South would be only 5 million). Throughout the Cotton Kingdom the despised Tariff of 1828 became known as the Tariff of Abominations. However, in terms of congressional action there was nothing the South could do about it.

On large plantations, slavery made cotton production viable on a massive scale. Most northerners had no *moral* objection to slavery in the South. Their prime concern was that free slave labor could lower the wages of northern freeman labor. One historian noted:

> As the North industrialized and its population grew, aided by a major influx of European immigrants, the northern majority in the House of Representatives also grew, making southern political leaders increasingly *[fearful]*. Southerners became concerned that they soon would find themselves at the mercy of a federal government in which they no longer had sufficient representation to protect their interests. . . . Senator Jefferson Davis from Mississippi stated that the northern majority in Congress made the government of the United States "an engine of northern aggrandizement" and that northern leaders had an agenda to "promote the industry of the *[North]* at the expense of the people of the South."

George McDuffie eventually would become governor of South Carolina. However, while he was in Congress he popularized the "Forty Bale" concept. It held that federal tariffs on cotton products had the effect of a thief stealing 40 bales of cotton from every 100 bales stored in southern barns. Although based upon fuzzy math, this premise was blamed for the decrease in cotton prices during the 1820s. During that decade southern farmers had witnessed the price of raw cotton drop from 18 cents a pound to only 9 cents a pound. Southern farmers realized that their livelihood and lifestyle slowly were being eroded away. Corrective action had to be taken.

Ordinance of Nullification: John C. Calhoun, a famous War Hawk from the South, had been elected vice-president of the United States in 1824. While serving in that office he authored the *South Carolina Exposition and Protest* in 1828. According to Calhoun, a state had the right to refuse to obey any federal law that the state deemed unconstitutional. Southern legislators took note but took no action at that time. However, four years later Congress passed an even more restrictive tariff stifling free trade. In response South Carolina passed a firebrand Ordinance of Nullification on 24 November 1832. The ordinance text read in part:

> We, therefore, the people of the State of South Carolina in convention assembled, do declare . . . that the several acts and parts of acts of the Congress of the United States . . . are unauthorized by the Constitution of the United States and violate the true meaning and intent thereof and are null, void, and no law, nor binding upon this State . . . .
>
> . . . We will not submit to the application of force on the part of the federal government . . . and *[we will resist]* any military or naval force against the State of South Carolina. . . . *[If the federal government uses military force against South Carolina]* the people of this state will henceforth hold themselves absolved from all further obligation to maintain or preserve their political connection with the people of other States, and will forthwith proceed to organize a separate government and do all other acts and things which sovereign and independent States may of right do . . . .

We will not submit to the application of force! Form a separate government! A sovereign and independent state! South Carolina had laid down the gauntlet. Although the Ordinance of Nullification was lengthy, the gist was simple and twofold. First, South Carolina declared the federal tariffs of 1828 and 1832 void and refused to obey them. Second, if the United States used military force in an attempt to mandate compliance with the tariffs, South Carolina would secede and form an independent government.

News of the revolutionary ordinance sent shockwaves throughout the nation's capital. Three months later in February 1833 the U.S. President signed a "Force Bill" that authorized use of naval and ground forces to militarily enforce tariff compliance in South Carolina. However, neither South Carolina nor other southern states were badgered into submission. They began plans to resist an expected naval blockade and military invasion. A compromise had to be found in order to avoid civil war.

Faced with secession by several more southern states, the federal government backed down. Congress capitulated and passed a new regulation providing for sharp reductions in the tariff percentage over a period of ten years. South Carolina responded by repealing its Ordinance of Nullification on 11 March 1833.

Both sides claimed a convoluted victory of sorts. Tariff rates had been lowered to a level satisfactory to southern states. On the other hand, in the eyes of the federal government, Congress had rejected the "states rights" concept of nullification. For the moment, secession and civil war had been avoided. Yet, the smouldering core conflict between North and South remained, and it would reach the boiling point 27 years later.

The Road to War: King Cotton continued to reign supreme in the South. By the close of the first 60 years of the 1800s the nullification crisis of the 1830s had been largely forgotten. Cotton had brought extraordinary wealth to farm families, and merchants shared their prosperity. Parentally funded schooling of children had become the norm. The future indeed looked bright. The seemingly boundless prosperity enjoyed by cotton producing states drew southerners into a tight economic and political alliance.

By mid-century in 1850 the United States had become united in name only. Two different and distinct regions had formed. States in the north, northeast, and mid-west enjoyed a booming economy fueled by mining, industry, and commerce. States in the south were driven by an agrarian economy and its primary cash crop, cotton. North and South effectively had evolved into two separate nations, a divided people with divergent aspirations. National politics had become sectional and pitted the two regions against each other.

By 1860 nationalistic fervor had swept through the southern United States. Recalling the Ordinance of Nullification in 1832, most southerners believed that secession from the oppressive United States would be in their best interests. Five primary factors led citizens in the South to this conclusion:

> First: Federal tariffs and taxes were rising again. Southerners, having almost no heavy industry of their own, resented tariff and tax increases on imports from Europe. Such increased costs served only to protect industries and jobs in the North.
>
> Second: Southern states had become outnumbered in the U.S. Senate and overwhelmed in the House of Representatives by the populous North. Consequently the South no longer had an effective voice in Congress.
>
> Third: The South had become rich by providing 68 percent of the entire world's raw cotton supply. Cotton production on large plantations was powered by slave labor, but abolitionist sentiment in the North was gradually rising.
>
> Fourth: To increase sales, newspapers in the North and South peddled outrageous embellished tales of regional controversy and conflict. Such irresponsible journalism further inflamed mistrust and passions on both sides.
>
> Fifth: Sovereignty, nationalism, and "states rights" became a mantra for the South. Exclusive of their common language, rural agricultural southerners had little in common with urban industrialized citizens in the North.

Tariffs enacted by Congress once again established increasingly high levies on finished cotton products imported from Europe. The 1860 election of Abraham Lincoln as U.S. President proved to be the final straw. His election without a single southern Electoral College vote confirmed that the South had lost all political influence at the national level. Worse yet, during his campaign for the presidency, Lincoln had called secession "treason."

Bad news and controversy sell newspapers. Since the days of Napoleon they always have, and they always will. Consequently, taking a cue from the European playbook, newspaper editors in the United States embellished the rift between North and South. This embellishment quickly grew into outright supposition, opinion, and fiction disguised as facts. Fanciful accounts of scandal, controversy, and immorality – replete with raucous inflammatory rhetoric – filled newspapers. Sales skyrocketed. Prompted by such irresponsible journalism, mistrust between North and South soared. That mistrust quickly turned into anger, and anger evolved into outright hatred.

> Both the North and South burned with righteous anger, because both passionately believed in the justice of their cause.
> *[John Miltimore, Causes of the Civil War, 2017]*

Mercenary newspaper rhetoric whipped up a religious fervor that was beyond debate in both regions of the United States. As national elections had neared in the autumn of 1860, southern editors painted presidential candidate Abraham Lincoln as a vile antichrist. Northern editors had countered by claiming that the envious South wanted to destroy northern industry and commerce. The hatred and false accusations on both sides became so bitter that they could not be reconciled. (The result would be mirrored 156 years later in 2016 when citizens would elect a wealthy businessman as U.S. President. His surprise election would be credited, in large part, to his denunciation of the 24-hours-per-day "fake news" industry and its thinly veiled political agenda.) John C. Calhoun, a fiery southern statesman, warned the South:

> If there be no protected power in the reserved right of the states, they must in the end be forced to rebel, or submit to have their paramount interests sacrificed, their domestic institutions subordinated, . . . and themselves and their children *[will be]* reduced to wretchedness.

Throughout the South calls for secession grew steadily louder. After all, southerners reasoned, their states had *voluntarily* joined the union known as the United States of America. Nothing in the U.S. Constitution prevented a state from voluntarily leaving that union. There were many obvious advantages in seceding, and few benefits in remaining. True, there were isolated calls for moderation. Yet, these cautionary voices were drowned out by a vast majority who favored withdrawal from the United States.

In both the North and the South, regional tension and hatred had become palpable. For the South the status quo became intolerable. Few could foresee the horror and tragedy lying ahead.

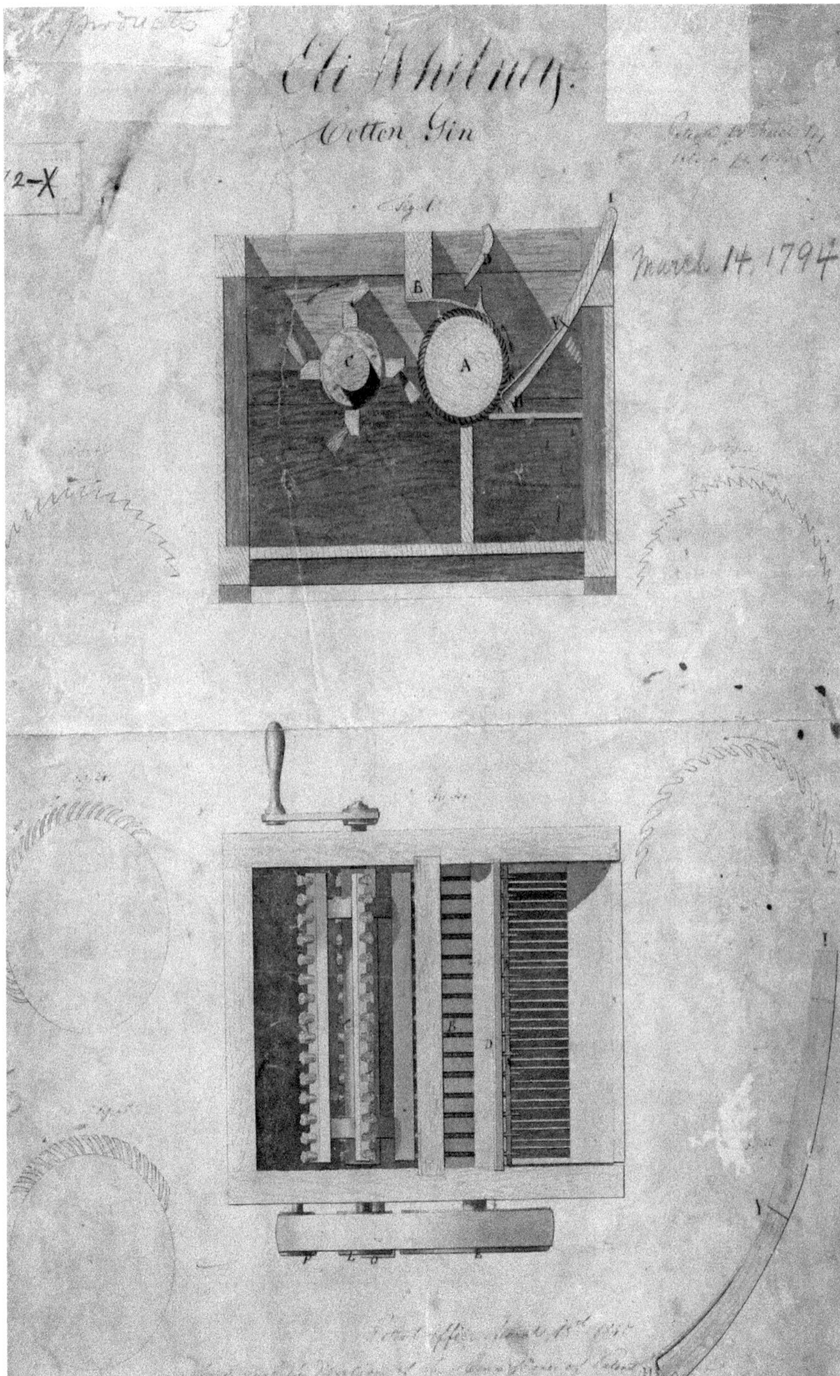

Eli Whitney's original cotton gin patent dated 14 March 1794.

John C. Calhoun, 7th vice-president of the United States, war hawk, and author of the South Carolina Exposition and Protest (photo courtesy of Library of Congress).

A cotton plant ready to be "picked" (photo courtesy of Department of Agriculture).

The notion that slave labor for cotton fields caused the Civil War has been reinforced by textbooks and fictional narratives for more than a century. Historians, however, argue for a more nuanced understanding. The Civil War was fought for many reasons, not solely, or even primarily, because of cotton on southern farms.

[Bruce Chadwick, historian, in his treatise, Causing the Civil War]

The War for Southern Independence

> We, the people of the Confederate States, each state acting in its sovereign and independent character, in order to form a permanent federal government, establish justice, insure domestic tranquility, and secure the blessings of liberty to ourselves and our posterity – invoking the favor and guidance of Almighty God – do ordain and establish this Constitution for the Confederate States of America. *[preamble to the Constitution of the Confederate States of America, 11 March 1861]*

Introduction: The titanic struggle pitting the United States against the Confederate States was the second civil war in United States history. Most history books would simply call it the Civil War, but it was more definitively known by other names. Northerners often called it the (1) War of the Rebellion, or (2) Second Revolutionary War, or (3) War Between the States. In the South the agonizing struggle was known as the (1) War for Southern Independence, or (2) War of Northern Aggression, or (3) War of Secession.

By whatever name the American Civil War remains the most deadly conflict in United States history. It decimated southern states. The entire fabric of southern society and culture was ripped asunder. The infrastructure and economy were crippled. Thousands of books have detailed military events during the brutal war, but those details will not be regurgitated herein. This chapter and following chapters focus upon the war's impact on combatants, their families, life in the South, and eventual reconciliation.

Secession: The first region to act was the firebrand state, South Carolina. Her legislature called for each county to send delegates to a Secession Convention that would be held in Charleston.

On 22 November 1860 in western South Carolina, Abbeville County began the secession process. Citizens and militiamen met in the Town of Abbeville and voted unanimously in favor of secession. They then passed a resolution demanding withdrawal from the United States. Many years later at the intersection of Magazine Street and

Secession Avenue in the Town of Abbeville a monument would be inscribed in part:

> On the hillside in the rear of this memorial on November 22, 1860, the first organized secession meeting was held.
> On that day the ancient artillery company, the Southern Rights Dragoons, and companies of minute-men from Abbeville, Greenwood, Cokesbury, Ninety-Six, Bradley, Due West, Donalds, Wickliffe, and Calhoun's Mill marched in line together with an immense concourse of loyal citizens, retiring to the grove. They there announced their intention of defending their Sovereign Rights as guaranteed *[to]* them in the Constitution of the United States.
> Resolutions demanding the immediate withdrawal of the state from the Union were unanimously passed, and representatives to the convention called by the legislature were nominated.
> And, thus, secession had its birth.

Other counties followed Abbeville's lead, and delegates to the state convention met in Charleston. On 20 December 1860 they approved an Ordinance of Secession by a 169-to-0 vote. Complete text of the brief ordinance follows:

> An Ordinance to dissolve the union between the State of South Carolina and other States united with her under the compact entitled, "The Constitution of the United States of America."
>
> We, the people of the State of South Carolina in Convention assembled, do declare and ordain, and it is hereby declared and ordained, that the ordinance adopted by us in Convention on the twenty-third day of May, in the year of our Lord one thousand seven hundred and eighty-eight, whereby the Constitution of the United States of America was ratified, and also all acts and parts of acts of the General Assembly of this State ratifying amendments of the said Constitution, are hereby repealed; and that the union now subsisting between South Carolina and other States, under the name of the "United States of America," is hereby dissolved.
>
> Done at Charleston the twentieth day of

> December, in the year of our Lord one thousand eight hundred and sixty.

Citizens celebrated in the new Republic of South Carolina. King Cotton had made them wealthy and was presumed to be a guarantor of their safety and independence. Cotton from southern states powered the textile empires of Europe. Southerners believed their role in the worldwide cotton industry would prompt international recognition, and military aid if necessary, from Europe.

South Carolina did not expect war with the United States, but as a cautionary measure the new republic ordered its militiamen to defend the seaport at Charleston. Rural Edgefield County wasted no time. On a clear Sunday morning 86 men of the Edgefield Riflemen militia got ready to depart for Charleston. Edgefield held an elaborate ceremony on the village square, and the *Edgefield Advertiser* newspaper would report three days later:

> THE BOYS ARE GONE
>
> The very noblest company of young men we have ever seen left this village on Sunday last for Charleston. . . . Scarcely ever was there such a scene of sympathy and fervor in our village as upon the morning of the march of these high-spirited young soldiers. There were no dry eyes among *[men and women of both races]* in the place. The Riflemen deeply felt the impressive occasion and went forth from amongst us well knowing what was before them. . . . We have plenty more soldiers to go *[to Charleston]* from Edgefield. Two more fine companies will be along immediately.

Speeches were made, women cheered and waved handkerchiefs, and a minister prayed for divine protection. The militia company then marched south to board a train bound for Charleston. The next day another company from Edgefield County, the Cherokee Pond Volunteers, followed them. Within two more days the Hamburg Minutemen also entrained for Charleston.

Within three days 286 militiamen from Edgefield County – an assortment of farmers, merchants, mechanics, day laborers, and artisans – had left their homes to defend their new republic. Within the year over 1,700 soldiers from that rural county would volunteer to defend their families, their homes, and their independence.

Genesis of the Confederate States of America: During the next two months, January and February, six more southern states (Georgia, Alabama, Mississippi, Louisiana, Texas, and Florida) seceded and became independent republics. These new sovereign entities each sent delegates to a convention in Montgomery, Alabama, in February 1861. They formed a new union, the Confederate States of America. Borrowing wording from the U.S. Constitution, they drafted a Confederate States Constitution. The document includes:

> Congress shall make no law respecting an establishment of religion, or prohibiting the free exercise thereof; or abridging the freedom of speech . . . .
>
> A well-regulated militia being necessary to the security of a free State, the right of the people to keep and bear arms shall not be infringed.
>
> No soldier shall, in time of peace, be quartered in any house without the consent of the owner; nor in time of war, but in a manner to be prescribed by law.
>
> *[Article I, Section 9, (12-14), the Constitution of the Confederate States of America, 11 March 1861]*

Shortly thereafter four more states (North Carolina, Virginia, Tennessee, and Arkansas) seceded and joined the Confederacy. The new nation established Richmond, Virginia, as its capital city on 30 May 1861. The Confederate president, Jefferson Davis, was a West Point graduate and former U.S. Secretary of War (later, *Secretary of Defense*). He called for 100,000 state militiamen to be transferred into the new C.S. Army. In that capacity they would stand ready to defend the Confederacy against a possible military attack.

Equipment and supplies carried by each new Confederate soldier were not uniform. Yet, in general each infantryman carried about seven pounds of ammunition, a rifle or a musket, a bayonet and scabbard, a canteen, and a blanket or shelter-half. Each also had been issued a haversack in which he carried his ammunition, rations, spare clothing, and personal effects. Some had a uniform, but many wore civilian attire. Pay usually was prompt. A private was entitled to eleven dollars per month payable in new Confederate currency.

In the decimated United States before Abraham Lincoln took

office the sitting president, James Buchanan, declared the secessions to be "invalid." Nonetheless he took no action and stated that the U.S. Constitution gave him no authority to do so.

Were the Secessions Legal under United States Law?: President Buchanan was correct. The Constitution did not address the issue. Each state that had banded together with other states in a union known as the United States of America had done so voluntarily. No external factors had legally compelled them to join the union. Now, once a state had voluntarily joined, could its legislature execute an ordinance of secession and voluntarily leave? The U.S. Constitution, Article IV, Section 3, Clause 1 (known as the Admissions Clause), is quoted below in its entirety:

> New states may be admitted by the Congress into this union, but no new states shall be formed or erected within the jurisdiction of any other state; nor any state be formed by the junction of two or more states, or parts of states, without the consent of the legislatures of the states concerned as well as the Congress.

This clause sets forth criteria under which Congress may admit a state into the union. There is no mention or reference to a state leaving the union. Legal scholars have long debated whether or not a state that had voluntarily joined could voluntarily leave. Some parties claim that the union is final and indissoluble. Others claim that sovereign states have an absolute irrevocable legal right to leave. For example, Texas seceded from Mexico in 1836 and became the Republic of Texas. Both sides claim a host of legal precedents to support their opinion. Yet, the crux of the matter is simple. One scholar cut to the heart of the debate when he wrote:

> The Constitution is totally silent on the question of whether or not a state may unilaterally leave, or secede from, the Union.

Tragically the debate would not be settled by a court system or by Congress or by state legislatures. Might would make right. The internationally accepted process would apply: Right of Conquest.

First Shots: The Confederacy had created a military department, and most state militiamen were administratively transferred to the new C.S. Army. In addition, on 16 March 1861 the C.S. Congress had established a C.S. Marine Corps and a C.S. Navy. Confederate troops occupied most former United States military installations in the Confederacy. The glaring exception was Fort Sumter.

Begun 32 years earlier in 1829, the Fort Sumter fortress sat atop a man-made island of 70,000 tons of New England granite that had been dumped into the sea in the entrance to Charleston Harbor. The massive masonry fortress bristled with cannons and was manned by 80 Union troops. When U.S. Navy warships approached to resupply the fort on 12 April 1861, Confederate batteries in and around Charleston began a 35-hour cannonade. Years later a large granite monument on nearby James Island would be inscribed in part:

> At half past four *[o'clock]* on the morning of April 12, 1861, from a mortar near this spot on Fort Johnson, the first shot of the War of Secession was fired by order of Captain George S. James of the Confederate States Army. The shell exploded above Fort Sumter, then occupied by Federal forces, thus beginning the tragic conflict which Americans endured for four years.

The Union garrison surrendered, and C.S. Army troops occupied and manned the bastion. Warfare had begun. The new United States president, Abraham Lincoln, issued a call for 75,000 volunteers to "put down the rebellion." He also ordered a naval blockade of southern seaports, and by mid-summer about 75 percent of southern imports and exports had been blocked.

Although European mercantile interests had huge stockpiles of raw cotton, the loss of cotton imports from the Confederacy caused cotton prices to soar. However, foreign military interference with the blockade would have constituted an act of war. Consequently neither Britain nor France would take any military action. The strangling blockade gradually became more and more effective, and within a year King Cotton would be of little help to the Confederacy.

Life in the Confederacy during the Summer of 1861: In many ways life in southern states continued as it had in the past. There was universal euphoria over the new Confederacy, but to a disinterested observer little else had changed. Schools continued their daily routines. Churches were filled each Sunday. Farmers who had not yet left their farm to join the C.S. Army planted and harvested their crops. They still devoted most fields to cotton production, for cotton bales could be stored for months or years. The impact of the naval blockade along the coast had not yet produced a significant adverse effect on the average citizen.

However, there were some changes. The Confederacy printed paper money. It was not backed by tangible assets, but instead by a simple promise to pay the bearer after the war. The Confederate dollar, called a *gray-back* because the reverse side was not printed, featured images of Confederate officials, gods and goddesses, and scenes of plantation life. The five-dollar note depicted five maidens representing (1) Justice, (2) Commerce, (3) Agriculture, (4) Industry, and (5) Liberty.

Postal service also changed. Contractors continued to carry mail from one post office to another, and postmasters initially accepted United States postage stamps. However, on 1 June 1861 the new Confederate States Post Office Department took control of the mail system and issued new Confederate States postage stamps. Each postmaster was required to sign a document pledging to "support the Constitution of the Confederate States of America."

> All laws of the United States of America in force and in use in the Confederate States of America on the first day of November *[1860]* and not inconsistent with the Constitution of the Confederate States . . . are hereby continued in force until altered or repealed by the *[C.S.]* Congress. *[Instructions to Post Masters adopted by the C.S. Congress, 9 February 1861]*

The First Land Battle: The war technically had begun on 12 April 1861 when Confederate troops fired on and captured Fort Sumter in Charleston Harbor. Thereafter, 500 miles to the north, Confederate troops formed a defensive line in northeastern Virginia just a few miles west of Washington, DC. They made no effort to advance beyond Virginia, the northernmost state in the Confederacy. There had been no further clashes with Union forces.

Lincoln searched for a way to void the secessions. He planned to militarily destroy the "rebel" army, march south, occupy the rebel capital at Richmond, and end the rebellion. At his direction the U.S. Army began massing 32,400 troops and artillery near Manassas Junction in northeastern Virginia about 15 miles west of Washington. Union and Confederate troops glared at each other across a small creek known as Bull Run. On the morning of 21 July 1861 throngs of sightseers and congressmen from Washington rode out in horse-drawn carriages to enjoy the military spectacle. Many had brought wine and picnic lunches.

The Union attack began with artillery fire against entrenched Confederate troops, and then Union infantry charged. Outnumbered, the gray-clad southerners fell back to Henry Hill, where they barely held their ground. Col. Thomas Jackson (later, *Stonewall* Jackson) earned his famous *nom de guerre* on the hill that day.

> There! There is Jackson, standing like a stone wall! Rally around the Virginians!
> *[Gen. Barnard Bee, CSA, rallying his men on Henry Hill, 21 July 1861]*

In the late afternoon Confederate reinforcements arrived by train, and the tide of battle began to turn. Then Col. J.E.B. "Jeb" Stuart and his cavalry showed up, and Stuart led a Confederate charge against a New York regiment. The Union line wavered. First, a few Union soldiers broke ranks and ran toward safety in the rear. Then Union companies began falling back in disarray. Panic gripped northern soldiers who remained in the fight, and within ten minutes entire Union regiments joined the wild and mindless retreat.

> Fleeing Union soldiers found the road *[to Washington]* jammed with carriages of congressmen and others who had ridden out to watch the fight.
> *[Manassas National Battlefield narrative]*

A total rout ensued. Northern newspapers would call it "The Great Skedaddle." All semblance of order fell apart. In a fanatical every-man-for-himself flight, Union soldiers ran toward Washington. Many threw down their guns to speed their mad dash for survival. Victorious Confederates took 1,312 prisoners and swarmed over abandoned artillery caissons, supply wagons, rifles, munitions, food, clothing, and assorted tools of war. Thus ended the First Battle of Bull Run (aka, *First Battle of Manassas*). The South had won, and had won decisively.

> The brilliant victory achieved by the Confederate forces at Manassas over the ruthless enemies of public liberty and constitutional law is an occasion of devout thanksgiving to Almighty God, and the citizens of Richmond reverently acknowledge His goodness and mercy.
> *[Richmond Dispatch newspaper, 23 July 1863]*

Although the South had won, it had suffered 2,032 casualties. Civilian doctors summoned to the battlefield were overwhelmed by the many hundreds of wounded C.S. Army and U.S. Army soldiers who still lay on the field. The first taste of battle had been sobering for the troops involved. Pvt. Thaddeus Freeman, CSA, had survived and was not wounded. He had enlisted on 12 June 1861, and only six weeks later he had been through his first battle. Two days thereafter he posted a letter to his family and wrote about soldiers they knew:

> Manassas, Va. July 23, 1861
> Dear Father, Mother, Brother, and Sisters,
> . . . We had a hard fight; the battle lasted seven hours . . . Jesse Stone *[was]* wounded, Wesley Jennings *[was]* shot in the head, . . . a cannon ball knocked Pickney Briggs down, . . . M. Rochelle was shot in the abdomen . . . Lieut. Johnson was killed dead, Col. Hampton was wounded slightly, and many others I will not mention. . . . I never expect to get back home alive.

Despite his prediction, Freeman would survive the war. Three years later he would be captured near Fort Harrison, Virginia, on 29 September 1864. Many Confederate prisoners died in captivity, but because he was a Mason, Freeman would be offered preferential treatment. If he swore allegiance to the United States and renounced the Confederate cause, he would be released and given a U.S. Government job with a salary of 60 dollars per month. Freeman would respond to that written offer with a written reply:

> I would rather die in prison than dishonor the cause for which I have fought.

A Golden Opportunity Slips Away: Prior to the battle the Union Army had not prepared defensive positions south and west of Washington or inside the city. Such precautions had been deemed unnecessary. Now fleeing Union troops had no forts, earthworks, or other defenses to protect them. Fearing that the Confederate Army would follow them and attack Washington, Union soldiers deserted the city and headed for Baltimore and other locations to the north. Washington was militarily abandoned.

In retrospect, in later years most historians would agree that Confederate commanders made a tragic mistake. The war would have ended by default if the C.S. Army had marched into Washington and occupied the defenseless capital city. There had been nothing to stop Confederates from doing so. Yet, they were content to remain in Virginia to "defend the homeland." Thus this golden opportunity quietly slipped away. It would not come again. Given this reprieve, the North would have time to regroup and devote its massive industrial strength and manpower advantage to a protracted war of attrition. The resulting bloodbath would devastate the South. It also would snuff out the lives of about 741,000 soldiers and would leave additional hundreds-of-thousands maimed and crippled.

> Before leaving the hospital I went into the garden and was shocked to see, lying about it, hands and arms and legs that had been amputated – a dreadful sight.
> *[Sgt. Berry Benson, CSA, diary entry circa 1864]*

Trouble on the Home Front: In communities, towns, and cities throughout the Confederacy news of the great victory at Bull Run (aka, *Manassas)* prompted even more enthusiasm for The Cause. Yet, as the war wore on this enthusiasm was tempered by realization that life had begun to change. Newspapers printed casualty lists that had been sent via telegraph to southern cities. Families first heard of the loss of sons, fathers, neighbors, and relatives from such lists. The Union naval blockade was increasingly effective, so finished goods and industrial products were no longer available. Small farms were the backbone of the South. However, most men of fighting age had joined the C.S. Army, so agricultural work and infrastructure support fell upon shoulders of elderly men, women, and small children. In cities dire food shortages became a way of life.

During winter months Confederate soldiers in the field suffered horribly from the cold. The South had no available source of wool, so families mailed all woolen clothing and blankets to sons, fathers, brothers, and uncles in the army. With no remaining wool blankets at home, families used their carpets as bed coverings.

The Confederacy printed paper money and bonds, but there was no silver or gold to back them up. The value of Confederate money began to nosedive, and most families eventually resorted to bartering in lieu of monetary transactions. Throughout the South progressive troubles mounted for the out-numbered, out-financed, and under-industrialized Confederacy. Southerners on the home front found the war to be an uninvited and inescapable presence in their lives. The first military draft in American history had begun on 16 April 1862. All Caucasian males aged 18 to 35 were eligible to be drafted into the C.S. Army. As the war continued the age range would expand.

Union Strategy vs Confederate Strategy: Union military strategists devised the Anaconda Plan. This military and economic scheme called for the U.S. Navy to strengthen the blockade of southern seaports and strangle the Confederate economy. The navy also would seize control of the Mississippi River and split Texas and Arkansas away from other southern states. By land and by sea, Union forces would block all shipments of heavy industry necessities to the Confederacy. This would cripple southern railroads, rolling stock, munitions factories, and other industries. The U.S. Army would keep military pressure on border states. Militarily and economically hamstrung, the crippled and outnumbered Confederacy would be unable to continue the war.

The Confederacy embraced a different philosophy, for its leaders knew they could never militarily defeat the much larger, far more populous, and heavily industrialized Union. They planned to gain diplomatic recognition from European powers. Such recognition was crucial, for the Confederacy needed to export cotton and food to Britain and France. In return the Confederacy needed to import industrial goods, medical supplies, and munitions. Warships in the U.S. Navy blockade would never fire on British or French merchant ships, the South reasoned.

On land the C.S. Army would defend the Confederacy and inflict so much damage and so many casualties on Union invaders that Union citizens would tire of the carnage. The Union then would have to recognize the Confederacy as a sovereign nation and stop the war. Consequently, for the Confederacy a tie was as good as a win.

Great Weakness of the Confederacy: The greatest weakness of the Confederacy turned out to be its transportation system. Troops, food, munitions, and all the tools of war moved via railroads. The North had four times more track mileage than the South. Yet, for the South the problem was not track mileage. It was track *gauge.* Each locomotive and railroad car has fixed axles, so they can operate only on tracks of a given width, or gauge. The Confederate rail system had been designed to connect with freight traffic on rivers, so southern railroads were a hodgepodge of different gauges. Different gauge tracks cannot be connected. After a train reached an intermediate city, military freight had to be off-loaded, re-loaded onto wagons, pulled by mules to a different railroad line, and re-loaded onto rail cars of a different gauge. At following junctions the process would have to be repeated unless the gauge chanced to be the same.

As the war progressed railroad lines and rolling stock in the Confederacy became overused, burned out, dilapidated, and used up. Before the war the South had gotten railroad equipment from rolling mills, factories, and machine shops in the North. The war cut off that source, and the U.S. Navy blockade prevented importation of new equipment. The Confederacy had no way to repair or replace worn out locomotives, freight cars, trestles, and bridges.

As the southern rail system deteriorated due to heavy use and Union sabotage, little could be done. The Confederacy did what it could, and it even ripped up entire rail lines in Florida and other remote areas for use as spare parts. By late 1864 the entire southern rail network would be more or less inoperable. Military strategists know the adage: "Field-grade officers study tactics, but general-grade officers study logistics and strategic initiative." Deterioration of the rail system turned out to be the Achilles' heel of the Confederacy.

Great Strength of the Confederacy: In Union states, public support for the war was divided. About 70 percent of Union citizens felt that if southern states wanted to leave the union, they should be able to do so. Good riddance, many northerners believed.

Conversely, citizens in the South were overwhelmingly united. They viewed southern independence with religious fervor and were willing to fight and sacrifice to achieve that goal. Southerners cherished ideals of gallantry, chivalry, and selfless service. Indeed, the greatest strength of the Confederacy was its famed armed forces. Robert E. Lee, "Stonewall" Jackson, and "Jeb" Stuart became living legends and were revered throughout the South.

Confederate military leaders became daily fodder in European newspapers of the post-Napoleonic era. Fanciful etchings of Gen. Jackson sucking on lemons in battle, Gen. Lee kneeling to pray for his soldiers, and Gen. Stuart leading a cavalry charge with plumes in his hat captivated European fancy. Confederate military commanders became larger-than-life heroes in the eyes of most Europeans.

Yet, the greatest military hero of the South was the individual Confederate soldier. He was ill-armed, ill-clothed, underfed, and often barefooted. Yet, he loyally bore the brunt of the fighting, suffering, privation, and dying. Such selfless dedication to cause and country would be exemplified by C.S. Army soldiers who crewed the Confederate submarine *H.L. Hunley*. Because of the experimental nature of the vessel only volunteers were accepted for its crew:

> 12 August 1863: In Charleston Harbor on a training mission, five C.S. Army volunteers drowned when *Hunley* sank.
>
> 15 October 1863: After the *Hunley* was raised a new crew of eight volunteers stepped forward for another training attempt. They all drowned when *Hunley* sank again.
>
> 17 February 1864: After *Hunley* once again was raised, eight more volunteers tried again. They successfully sank the blockading warship *USS Housatonic*. However, *Hunley* also sank, and its third crew of volunteers all drowned.
>
> 19 February 1864: The C.S. Army expected the submarine to be found, raised, and refitted for another mission. Within two days 86 more C.S. Army soldiers volunteered for the next sortie, but *Hunley* would remain missing for 136 years.

Flags of the Confederacy: Background information is essential. A half-century earlier in 1810 the short-lived Republic of West Florida had created a national flag. It featured a solid blue field emblazoned with a large five-pointed white star. Twenty-six years later in 1836 this same flag design was adopted by the Republic of Texas, and over passing years Texas would become known as the Lone Star state.

Mississippi had seceded from the United States in January 1861 and had begun flying a new flag atop its capitol dome. This flag, almost identical to the Texas flag, featured a blue field and a single large white star. As a tribute to this banner, Harry McCarthy wrote the upbeat and peppy marching song, *The Bonnie Blue Flag*. One of the many choruses begins:

> Hurrah! Hurrah!
> Hurrah for the Bonnie Blue Flag!
> That bears a single star!

Both the song and the flag became wildly popular, and the new Bonnie Blue Flag became the *unofficial* flag of the Confederacy.

Another unofficial flag debuted in December 1861. The Army of Northern Virginia designed a square battle flag. The flag featured a diagonal blue Saint Andrews Cross atop a red field. White stars within the blue cross corresponded to the number of states in the Confederacy at the time. Various versions of this battle flag were flown by most Confederate military units throughout the war.

The first *official* national flag of the Confederacy also debuted in 1861 and was adopted by the C.S. Congress. Nicola Marschall, a Prussian artist, designed the flag, which resembled the flag of the Austrian Empire. The new flag featured two horizontal red stripes, or bars, with one identically sized white stripe between them. In the upper left corner a blue canton was emblazoned with small white stars arranged in a circle. There was one star for each state, and stars were added as more states joined the Confederacy.

Known as the Stars and Bars, this flag had been carried into battle for the first time at Bull Run (aka, *Manassas*) on 21 July 1861. At Bull Run and at battles thereafter the flag fell into disfavor. Amid dense smoke on battlefields, flags were hard to see under the best of circumstances. Yet, soldiers had to "follow their colors." Both the new Confederate flag and the United States flag had the same colors:

red, white, and blue. If the Stars and Bars became wrapped around the flagstaff or hung limp, it looked exactly like the United States flag, the Stars and Stripes. In battle this caused mass confusion and presented a huge problem. Writing in *Southern Literary Messenger* in 1862, George Bagby noted:

> The present *[flag]* is universally hated. It resembles the Yankee flag, and that is enough to make it unutterably detestable.

The C.S. Congress took action to remedy this issue in May 1863. The Congress adopted a new national flag and named it the Stainless Banner. The canton consisted of the small battle flag of the Army of Northern Virginia. This small canton was superimposed on a solid white field, and white was chosen as a symbol of purity.

The new flag was despised by Confederate armed forces. When it hung limp on the flagstaff, often only the solid white field was visible. In such instances it looked like a flag of truce, a signal of surrender. Further, the solid white field did not remain white for long, for it quickly became heavily soiled under field conditions. One newspaper editor claimed that after only a month of field usage the white field would resemble a "dirty gray rag."

Late during the war in March 1865 the C.S. Congress adopted an updated national flag, the Blood Stained Banner. It was identical to the Stainless Banner with one exception. A broad red stripe extended along the right edge of the white union. The war ended the month after this flag was adopted, so few were produced.

During the war Confederate ships, forts, batteries, and troops flew a variety of flags. Confederate forces in the west often used the Van Dorn battle flag, which featured a white crescent and white stars. First Corps, Army of Tennessee, flew a blue battle flag with a red Saint George's Cross. The 1st Cherokee Mounted Rifles flew a battle flag almost identical to the Stars and Bars, but the words "Cherokee Braves" were included on the white stripe.

The most famous Confederate flags were the Bonnie Blue Flag and the Army of Northern Virginia battle flag. In addition most regiments had their own regimental colors. Also, individual state flags were frequently borne into battle and displayed during dress parades. Alabama and Texas each had two state flags, and South

Carolina had three.

Seal of the Confederate States: Often called the Great Seal of the Confederacy, the seal was designed to authenticate documents issued by the Confederate government. The C.S. Secretary of State arranged for manufacture of the seal in London.

The seal features George Washington on horseback. Washington was chosen as a model because of his personal character, political leadership, and military acumen. On the seal Washington's image is surrounded by a wreath constructed of agricultural products of the Confederacy: tobacco, corn, wheat, rice, cotton, and sugarcane. Near the top margin of the seal are the words: "The Confederate States of America: 22 February 1862." That date reflects the start of C.S. Government in Richmond after the general election when Jefferson Davis was inaugurated as president. The date also is the birthday of George Washington.

Near the seal bottom margin are the words Deo Vindice, the national motto of the Confederate States. Translated from Latin, the words mean "God is our vindicator" or "God will vindicate us." This wording was not chosen lightly. The C.S. Congress noted that the U.S. Constitution is silent on the subject of the Deity. Accordingly the C.S. Congress stressed that the Confederacy was established under God's governance. The brief preamble to the Confederate constitution includes the clause: ". . . invoking the favor and guidance of Almighty God."

Design of the new seal was not completed until April 1863, and the Confederacy ordered embossing dies from a London engraver. Via blockade running ships, the silver dies safely reached Richmond. Yet, because of dangers in running the blockade, the accompanying embossing press had been shipped only as far as Bermuda. It never reached Richmond.

A century after the war the silver dies would be displayed in the Museum of the Confederacy at Richmond. The embossing press and brass replica dies would be displayed in the National Trust Museum at St. Georges, Bermuda.

The Noose Tightens: Foreign diplomatic recognition and intervention never came. The noose of the strangling naval blockade grew ever tighter and could not be broken. Southern infrastructure slowly withered and died. Yet, the out-manned and out-gunned Confederacy continued to fight. Seeking a decisive victory that might end the war, the C.S. Army had invaded Maryland in 1862, then Pennsylvania the next year. On 1-3 July 1863, Union and Confederate armies battled in hills and fields around Gettysburg, Pennsylvania. Of the 163,000 combatants, 51,000 became casualties. The bloody battle was a tactical draw. Yet, running out of munitions with no hope of resupply, the South had to abandon its invasion and return to Virginia. On 7 July, four days after the fighting ended, Pvt. Alex McNeill, CSA, posted a letter to his wife:

> My Dear Wife,
>
> I am thankful to an ever kind and merciful God that I have again been spared. . . . The enemy opened upon us with grape and cannister shot, and our brave boys were mowed down by the score. . . . You could constantly see men falling on all sides, and the terrible missiles of death were flying thick and fast everywhere . . . . We went right up under the enemy's guns, but our ranks were so decimated that we found it impossible to carry the position. In our regiment we carried about 350 men into action and lost in killed and wounded, 181. . . . We buried *[our dead]* as best we could in the same grave on the battlefield and placed a head-board at each one's head, with his name and Company and Regiment.

As late as autumn in 1864 the Confederacy was still winning military victories, but in this war of attrition the agricultural South could not outlast the industrialized North. Slowly the war took a toll on C.S. Army soldiers. In a letter to his parents a Confederate private wrote about the loss of family members and friends:

> . . . There were the three Thompkins brothers, and little Jim, the Nephew, as we called him. There were not four braver men in the company, for I stood and fought side by side with them. A friend in battle, a friend in camp, was a friend indeed. I remember well the night after the battle when I went to Gus for consolation, telling him how badly Brother Presley was wounded and that Brother Oscar was missing,

> and his reply was that I *[should]* console myself, that Jack *[also]* was missing, and such was the fate of War.

Another Confederate private described the chaos of a nighttime skirmish in a letter to his family:

> . . . We charged their works. We were so close that the flash light from *[their]* guns blinded us. That is when I saw your father. He fell at first fire, and I thought he was killed, but *[he]* rose and turned and ran by me; so did I and everyone that had not been shot down.

Morale sagged on the home front. There was no money. There was little food. Citizens knew that men from every community had been killed or maimed or had died of camp diseases. Infrastructure had fallen apart. The earlier prospect of military victory had dimmed. In response the Confederacy took various steps in an effort to bolster public support and morale. The C.S. Congress published an upbeat white-paper designed for newspaper publication. In part it states:

> What *[our Confederacy]* has accomplished, with a population so inferior in numbers and means so vastly disproportionate, has excited the astonishment and admiration of the world. The war in which we are engaged was wickedly and against all our protests . . . forced upon us.
>
> Our army is no hireling soldiery. It comes not from paupers, criminals, or emigrants. It was raised by the free unconstrained unpurchasable assent of the men.
>
> . . . *[Our]* houses are pillaged and burned *[by the invading U.S. Army]*. Churches are defaced. Towns are ransacked. Jewelry and mementos of the dead are stolen. . . . History will hereafter pillory those who committed and encouraged such crimes in immortal infamy.

By January 1865 the Confederacy was a thin shadow of what it had been four years earlier. Most farms had been abandoned. Horses, mules, hogs, and cattle were gone. With no maintenance for four years, roads were impassable or in a state of disrepair. Bridges and ferries had been abandoned or washed away. Large cities like Atlanta had been burned, and the same fate soon would befall Columbia to the east. Shops and warehouses lay in ruins. Crucial river traffic was at a standstill. Confederate money was worthless. Life savings invested in slaves soon would be worthless also.

Futile Peace Conference: In the North, Lincoln and his generals looked for a way to end the carnage. In the South, Confederate officials searched for a way to avoid total military capitulation and the death of their Confederacy. After contacts by intermediaries the two governments met in Union-occupied Norfolk, Virginia, on 3 February 1865. Lincoln and his secretary of state, William Seward, conferred with the Confederate vice-president, Alexander Stephens, and his two aides. The Confederacy proposed a package deal with three elements:

First: Abolition of slavery, which would be replaced with a more mild, yet subordinate, status for servants.

Second: Union recognition of the Confederacy and cessation of all hostilities.

Third: A joint Union-Confederate attack on Mexico to unseat despised dictator Maximilian.

Although the conference was civil, even cordial, no agreements were made. Consequently the last realistic hope for survival of the Confederacy silently slipped away.

Death of the Confederacy: By the spring of 1865 the C.S. Army tottered on the brink of collapse. Its troops had been forced into defensive redoubts. Munitions had almost run out. Food was scarce. Petersburg fell on 2 April 1865, and Richmond had to be evacuated the next day. A week later on 9 April with supply lines severed and with no remaining tools of war with which to continue the struggle, Gen. Robert E. Lee surrendered his entire Army of Northern Virginia near Appomattox Court House. Commanders of other Confederate armies would follow his lead. The required Amnesty Oath, initially given verbally, later would be documented in writing:

AMNESTY OATH

I, *(printed name)*, of *(home address)*, do solemnly swear, in the presence of Almighty God, that I will henceforth faithfully support, protect, and defend the Constitution of the United States and the Union of States

> thereunder, and that I will in like manner abide by and faithfully support all laws and proclamations which have been made during the existing rebellion with reference to the emancipation of slaves, so help me God.
>
> ______*(signature)*______

Jefferson Davis, the Confederate president, had left Richmond during its evacuation and had headed southwest toward Texas. Once there he planned to reorganize the Confederate government and continue the struggle. In his wagon train were the Secretary of State, Secretary of War, Secretary of the Navy, Postmaster General, and six general-grade officers. Davis and his entourage reached Abbeville, South Carolina, on 2 May 1865.

The first secession meeting had been held in Abbeville over four years earlier on 22 November 1860. Now the Confederacy would be dissolved in the town, giving Abbeville the moniker, "Birthplace and Deathbed of the Confederacy." Erected years later, a granite monument in the roadway median at the junction of Main Street and Greenville Street in Abbeville would explain:

This tablet was placed here
to commemorate
the last meeting of the
CONFEDERATE STATES CABINET,
which was held in the Burt house nearby,
directly in front of this stone.
The following cabinet members were present:

Jefferson Davis	President
Judah P. Benjamin	Secretary of State
John C. Breckinridge	Secretary of War
S.R. Mallory	Secretary of the Navy
John H. Reagan	Postmaster General

A Council of War was held at the same
time with the Cabinet and the following
generals in attendance:

W.C. Breckinridge	Geo. C. Dibrell
Basil W. Duke	S.W. Ferguson
J.C. Vaughn	Braxton Bragg

It was decided after deliberation

and discussion that it was useless to
continue the war longer and that the
Government should be disbanded.
M.H. Clark, Acting Treasurer, Confederate
States of America, says: "The last Cabinet
meeting, which could be called such, was
held at Abbeville on the 2nd day of May, 1865."
A full history of these events may
be found in the Office of the Clerk of
Court in Abbeville County.
The Confederate States
of America
22 February 1862
Deo Vindice

Jefferson Davis and his entourage were captured by Union troops on 10 May 1865 near Irwinville in rural Georgia. The Confederacy died quietly and faded away into the history books.

Fourteen years later in 1879, eighty miles east of Abbeville in the City of Columbia, a large memorial for Confederate soldiers killed during the war was erected in front of the South Carolina State House. The inscription on its southern side is quoted below:

Let the stranger,
who may in future times
read this inscription,
recognize that these were men
whom power could not corrupt,
whom death could not terrify,
whom defeat could not dishonor.
And let their virtues plead
for just judgement
of the cause in which they perished.
Let the South Carolinian
of another generation
remember
that the state taught them
how to live and how to die,
and that from her broken fortunes
she has preserved for her children

the priceless treasure of their memories,
teaching all who may claim
the same birthright
that truth, courage, and patriotism
endure forever.

Epilogue: The brutal War for Southern Independence remains the most deadly war in American history. About 685,000 men (394,000 from the North and 291,000 from the South) had been killed in battle or died from wounds and disease. About 56,000 more had perished in squalid POW camps because of privation, starvation, and lack of medical care. In total about 741,000 men from the armed forces of the North and South perished during the war.

The terrible toll of war continued. In addition to those who died, 63,000 survivors had lost arms or legs. Additional hundreds-of-thousands who survived had been blinded, burned, maimed, crippled, or otherwise incapacitated. By the tens-of-thousands, civilians of all ages and both sexes had fallen prey to roving bands of guerrillas, partisans, and vigilantes. The blow to posterity was incalculable. Whatever the sums and effects of such irreparable losses may have been, they would multiply exponentially for generations to come (see the chapter, "Casualties of War").

On fame's eternal camping ground
Their silent tents are spread,
And glory guards, with solemn round,
The bivouac of the dead.
[Theodore O'Hara, The Bivouac of the Dead, 1847]

Oil painting depicting a fanciful image of the bombardment of Fort Sumter in Charleston Harbor on 12 April 1861 (image courtesy of Library of Congress).

Image of the Seal of the Confederate States. The seal image includes George Washington on horseback and the Confederate States motto, "Deo Vindice."

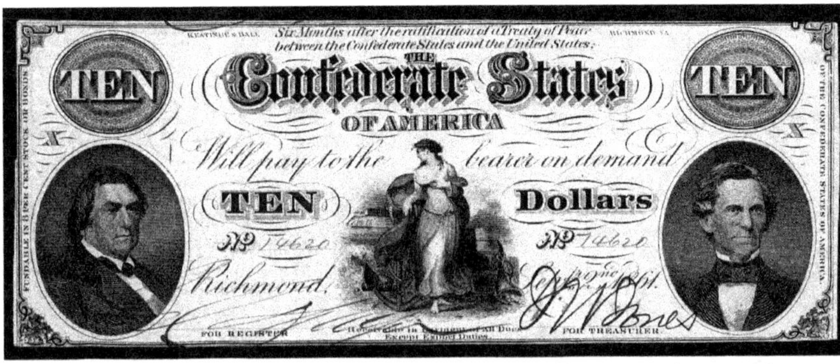

Image of a Confederate States ten dollar bill. Note the wording on the bill: "Six Months after the ratification of a Treaty of Peace between the Confederate States and the United States, the Confederate States of America will pay to the bearer on demand, Ten Dollars."

Etching depicting *<u>CSS Advance</u>*, *a swift Confederate Navy steamer used as a blockade runner (image courtesy of U.S. Navy).*

The "Bonnie Blue Flag" debuted in early 1861 and became the first unofficial flag of the Confederate States.

The "Stars and Bars" became the first official Confederate States national flag. It would be replaced in 1863 by a new national flag with a totally new design, the Stainless Banner.

A young Confederate States Army soldier, Pvt. Edwin F. Jemison, 2nd Louisiana Infantry Regiment. He was killed-in-action in July 1862 during the Battle of Malvern Hill when a solid cannon shot decapitated him.

Some of the Confederate dead laid out for burial at Spotsylvania, Virginia, on 18 May 1864.

Workmen recovering skeletal remains of Union soldiers killed-in-action at Cold Harbor, Virginia, many months after the bloody struggle there.

Confederate Memorial in front of the State House in Columbia, South Carolina (photo by the author circa 2010).

War loses a great deal of its romance after a soldier has seen his first battle.
[Col. John Mosby, CSA, Mosby's War Reminiscences, 1887]

Wounded and Sick Confederate Soldiers

> I saw the battle-corpses, myriads of them,
> And the white skeletons of young men – I saw them;
> I saw the debris, and debris of all the dead soldiers of the war.
> But I saw they were not as was thought,
> They themselves were fully at rest – they suffered not;
> The living remain'd and suffer'd . . . .
> *[Walt Whitman, When Lilacs Last in the Door-Yard Bloom'd, 1865]*

Nineteenth Century Warfare: At the beginning of the War for Southern Independence in 1861 the extent of medical knowledge was primitive. In Europe the pioneering work of Koch and Pasteur had barely begun. Blood typing and x-rays were many decades away. Antibiotics were not available. Doctors did not understand infection, and most of them had never treated a gunshot wound. No effort was made to maintain medical sterility. Military camp hygiene was exceptionally poor. Communicable diseases ran rampant, for filthy overcrowded camps made perfect disease breeding grounds. Soldiers ran a high risk of getting shot and killed in battle. However, they had a much greater risk of dying from disease or infection.

The effect of weapons on the human body were well understood. In Europe, Napoleon had referred to artillery as his King of Battle. New cannon shells were hollow cast iron spheres filled with black powder that would detonate, sending lethal shards of shrapnel ripping through the air in all directions. Arms and legs were scythed away. Bodies were torn apart. Fired from the same artillery, cannister shot consisted of dozens of small iron balls. Cannister made artillery the equivalent of an enormous shotgun, and it was extremely effective against attacking infantry or cavalry. Shrapnel or cannister wounds in a soldier's extremities were disabling at the least. Such wounds in his torso were almost always fatal. By the 1860s artillery had become a stone cold battlefield killer.

Smoothbore muskets had faded away, and new rifled weapons were vastly more accurate and deadly. The new Minie-ball had been developed 12 years earlier by a French officer, Claude-Etienne Minie.

It proved to be a devastating tool. The huge .58 caliber bullet was made of soft lead. This relatively slow-moving projectile flattens and expands on impact with human flesh. It creates a wound that grows ever larger as the bullet tears deeper into tissues, shattering bones above and below the point of impact.

Splintered bones and gaping wounds became the Minie-ball calling card. Because the soft lead continues to expand as it rips through flesh, it often does not exit the body. Arms or legs with shattered bones from such a bullet could be amputated, giving a wounded soldier a chance of survival. Conversely, a Minie-ball wound in one's trunk was usually a death sentence.

Military Medicine: During months before the carnage at Bull Run (aka, *Manassas*) neither the North nor the South had expected the war to last more than a few months. Little planning had addressed medical treatment for wounded soldiers. After the struggle at Bull Run some wounded men from both armies had lain on the battlefield, shrieking in agony, for two days before getting medical attention. Writing about such unattended battle casualties in another war many years later, a French historian would explain:

> The wounded were still lying there just like on the first day, intermingled with men who had died several days ago and were beginning to rot. They were lying there unattended in the sun, being eaten alive by the rats and vultures. If only they had all been dead!
> *[Dr. Bernard B. Fall, Street Without Joy, 1961]*

There had been no organized military ambulance system (an *ambulance* was a wagon pulled by one or two horses) in either army. Before the battle a few civilian contractors had been hired to use their own horses and wagons to evacuate men who would be wounded. Unfortunately, as soon as opposing artillerists began dueling these civilians had turned tail and vanished. Compounding the absence of a functional ambulance system, there was no organized system of battlefield medical aid. After the fighting ended civilian doctors had been summoned to the battlefield to do what they could for wounded men. This inspection and "first aid" to casualties took over 48 hours. Wounded Confederate soldiers plus captured and wounded Union soldiers who survived long enough to make the trip were eventually moved by wagon and rail to make-shift hospitals in Richmond.

After Bull Run the South was quick to establish a medical corps within its army. In fact, six months earlier the first step already had been taken when the Confederate government created the C.S. Army Medical Department:

> The Medical Department shall consist of one surgeon-general with the rank of colonel, four surgeons with the rank of major, and six assistant surgeons with the rank of captain . . . .
> *[an act of the C.S. Congress, 26 February 1861]*

Within a year the reality of protracted warfare made the initial authorization numerically obsolete. Vastly more medical personnel would be required in the C.S. Army. New authorizations provided:

– One surgeon general, rank of brigadier general.
– 1,000 surgeons, rank of major.
– 2,000 assistant-surgeons, rank of captain.
– Acting assistant-surgeons as needed, rank of lieutenant.

A small medical corps for the C.S. Navy had been authorized by the C.S. Congress on 16 March 1861. By April 1862 this staff had been enlarged to 22 surgeons, 15 qualified assistant-surgeons, and 30 assistant-surgeons without essential qualifications. In May 1863 the Confederate Congress empowered the president, Jefferson Davis, to appoint more naval medical personnel "as needed."

Causes of Death: The bloody War for Southern Independence was fought during an era often known as the Middle Ages of Medicine. Doctors labored admirably and did what they could during a time before the causes of disease and infection were understood. Sterile procedures and antisepsis were unknown. Few effective and safe medications were available. For every two soldiers killed in battle, three more died of disease.

Armies in both the North and South initially required a medical inspection of potential enlistees. The written requirements for such inspections were stringent. One requirement is quoted below:

> . . . The medical officer is to examine him stripped; to see that he has free use of all his limbs; that his chest is ample;

> that his hearing, vision, and speech are perfect; that he has no tumors or ulcerated or extensively cicatrized legs *[and]* no rupture of cutaneous affection; that he has not received any contusion or wound of the head that may impair his faculties; and has no infectious disorder or any other *[condition]* that may make him unfit for military service.

Wishful thinking! In reality the examining doctor rarely came close to meeting written requirements. As the war dragged on and the Confederacy became more desperate for replacement soldiers, the pre-induction examination became moot. If a recruit could walk, shoot a rifle, and had front teeth (*front teeth* were required to bite and rip open the cartridge wrapper for newer rifle cartridges) he was able to enlist. The medical inspection requirement became ignored to the extent that roughly 85 women, disguised and dressed as men, were able to enlist during the last 10 months of the war.

A military doctor in the C.S. Army was known as a *surgeon*. However, the terms doctor, surgeon, and physician were used loosely. The first so-called medical school in the United States had been established in Philadelphia. At that time there was no entrance exam, and no prior schooling was required. Those who paid to attend the two-year school were given lectures during eight months of the first year. During the second year the same lectures were repeated. There was no internship, no clinical experience, no licensing board.

Many military doctors were woefully unqualified. Most of them had never performed any type of surgical procedure, so they relied upon the *Military Surgery Manual* written by Dr. Samuel Gress. By mid-1864 almost anyone who claimed to be a doctor and had the requisite medical kit was accepted as a C.S. Army surgeon. Yet, in trying circumstances most of them performed admirably in the field. One surgeon, Dr. Daniel Holt, sent a letter to his wife:

> . . . At early dawn while you, I hope, were quietly sleeping, I was up at Surgeon's Call, and before breakfast prescribed *[medicine]* for 86 patients at the door of my tent. After *[breakfast]* I visited the *[field]* hospitals and a barn where our sick are lying and gave medicines and prescriptions to one hundred more. I had no dinner. . . . A cold bite was eaten *[for supper]*. After this, in the rain, I

> started for Sharpsburg, four miles distant, *[to try to get more]* medical supplies.

Forward Aid Stations: During 1861 there had been no organized method of removing wounded men from fields of battle. In early 1862 this task was assigned to members of regimental, division, and corps bands. By late 1862 the process had evolved. Unwounded infantrymen would carry wounded comrades to get medical care. Stretchers, when available, would be used for this purpose. As the war dragged on, designated stretcher bearers became an integral part of the C.S. Army Medical Corps.

By the end of 1862 the Confederate Army had developed a three-tier system of medical treatment for soldiers who were wounded in battle. Ideally, although not practical in many clashes, wounded men first would be treated at a forward aid station. There an assistant surgeon, hopefully trained in first aid, had only two goals: (1) stop the bleeding, and (2) treat or prevent shock. The usual remedy for intractable pain was liberal doses of whiskey. Medicinal painkillers morphine and opium could be used if they were available. Wounds would be packed full of lint and then tightly bandaged. Compressed lint hopefully would be sufficient to stop the bleeding.

If lint and a tight bandage would not stem the loss of blood, a tourniquet could be applied to arms or legs between the wound and the trunk. When cinched tight enough to stop all blood flow to the extremity, such a tourniquet could save a soldier's life. Depending upon the tide of battle, transportation by ambulance (a horse-drawn wagon) to more advanced medical treatment often took many hours. Sometimes it took over a day. Consequently, when a soldier with a tourniquet finally reached a field hospital the atrophied limb would have to be amputated.

> I notice a heap of amputated feet, arms, legs, and hands, a full load for a wagon. Dead bodies lie nearby . . . their names *[written]* on pieces of barrel-staves or broken boards, stuck in the dirt.
> *[Walt Whitman (then a U.S. Army nurse), near Fredericksburg, Virginia, 21 December 1862]*

Field Hospitals: After leaving a forward aid station a wounded soldier ideally would be moved by ambulance to a field hospital. The trip would be a rough bumpy wagon ride over rural terrain, an agonizing experience for severely wounded men. Field hospitals at the regimental level would be set up several miles from the field of battle, out of anticipated range of enemy artillery. Any home, barn, or store would suffice. After major battles hundreds of wounded men reached such hospitals. Upon arrival each casualty was assigned into one of three categories:

– Mortally wounded men.
– Grievously wounded men who could be saved.
– Wounded men who were not in imminent danger of dying.

The attending surgeon would begin by treating wounded men in the second category. A barn door supported by two barrels or sawhorses served as an operating table. The scene resembled a butcher shop. One wounded colonel described the field hospital in which he received treatment:

> When I was carried into the building and looked about I could not help comparing the surgeons to fiends. . . . All around on the ground lay the wounded men. Some of them were shrieking, some were cursing and swearing, and some were praying.

Surgeons and their assistants had no caps under which their hair could have been tucked. They did not wear surgical gowns. Unaware of the importance of sterility, neither surgeons nor their assistants washed their hands or their surgical tools before starting to treat each new casualty. The operating table was not cleaned after each patient was removed. One soldier from the 1st Virginia Cavalry wrote about gruesome amputations:

> Screaming victims were having their legs and arms cut off. Surgeons and their assistants, stripped to the waist and bespattered with blood, stood around, some holding the poor fellows while others, armed with long bloody knives and saws, cut and sawed away with frightful rapidity, throwing the mangled limbs onto a pile nearby. Men were soon overwhelmed by the prayers, the curses, the screams, the flies, and the sickening stench.

Surgeons used their fingers to probe wounds for shell fragments and Minie-balls that had not exited the body. However, the most common operative procedure was amputation of arms and legs. Surgeons earned the moniker, "sawbones." Because antibiotics had not then been discovered, experience had proven that amputation of injured limbs was the best way to prevent infection, gangrene, and death. About 75 percent of wounded men with an amputated limb would survive. Standard procedure called for severing the limb immediately above the wound and above any shattered part of the bone. The farther from the torso the amputation was performed, the greater the chance of survival.

> Shattering, splintering, and splitting of a long bone by a Minie-ball were frightful, and experience taught surgeons that amputation was the only means of saving life.
> *[regimental surgeon, Tennessee Volunteers, 1863]*

Amputations done within 12 hours of the injury resulted in the highest survival rate. Amputations performed over 48 hours after the injury caused the mortality rate to skyrocket. Once a surgeon and his assistants became experienced, an arm could be amputated in three to five minutes. Legs took up to ten minutes. Skin from the amputated limb would be removed. It would be placed over the raw surgical wound and stitched to the stump.

Contrary to popular belief, few amputations were performed without anesthesia, which had been introduced in the United States during the 1840s. Ether was commonly used, but chloroform was the most popular anesthetic. It was widely available, nonflammable, and simple to use. When properly administered it had a rapid onset of effectiveness but did not render a patient totally unconscious. Yet, sedated men felt no pain. For example, Gen. Thomas "Stonewall" Jackson was sedated with chloroform when his left arm was amputated by Dr. Hunter McGuire in May 1863. Several hours after the procedure Jackson told the doctor that he clearly heard the sound of the saw as it cut through the bone in his arm. However, he said that he felt no discomfort.

After surgeons at a field hospital plied their trade, slightly wounded soldiers convalesced there for several days. Then they returned to their unit and performed administrative duties until they fully recuperated. Grievously wounded men and amputees remained at a field hospital only until they had medically stabilized. Then they were sent to a general hospital in a major city.

From a field hospital, an ambulance carried seriously wounded men to the closest railroad line. Such a destination might be five miles away, or perhaps 50 miles away. The wagon trip might take an hour, or it might take more than a day. After arriving at the railroad, wounded men were loaded into boxcars for shipment to a general hospital. That trip often took over a day if multiple railroads of different gauges were involved. In the winter unheated boxcars were freezing cold. En route there was no heat, no food, no water, and nothing to ease the pain suffered by wounded men.

After arrival at a rail terminal near a general hospital, wounded soldiers were placed in wagons for the final leg of their journey. They would find that medical care in large military hospitals would range from adequate to excellent.

General Hospitals: The Confederacy used the term *general hospital* liberally during the war. Buildings used for treatment of sick and wounded soldiers ranged from private halls, to hotels, to warehouses, to factories, to college buildings, to large complexes built specifically for medical care. During the war the C.S. Army assigned the term *general hospital* to 39 buildings and enormous building complexes in Virginia. Elsewhere there were 4 such hospitals in Florida, 49 in Georgia, 3 in Mississippi, 21 in North Carolina, 12 in South Carolina, and 2 in Tennessee.

Richmond was the Confederate capital. Richmond also was the home of the Medical College of Virginia, the only medical school in southern states. Throughout the entire war about 58 percent of Confederate casualties were sustained within 80 miles of Richmond. More Confederate troops fought in Virginia than in any other state. Five railroads converged in Richmond, so most sick and wounded soldiers in the state could be transported to the city by rail. Richmond became a booming Confederate medical center.

Medical Dilemmas: In 1861 the general hospital concept had been in its infancy. When badly wounded soldiers became ambulatory the C.S. Army furloughed them so that they could go home to complete their recovery and regain their strength. The army quickly abolished this policy as new general hospitals were built, for they provided enough beds for convalescents. Also, hospital surgeons would be able to decide when a soldier was ready to rejoin his unit.

During early months of the war, surgeons had provided their own medical supplies. However, by 1862 the Confederacy supplied all regiments with medical packs that included both surgical instruments and medications. Medical packs were greatly prized, for the army had limited access to medicine. In particular, medical instruments were hard to obtain. The Confederacy got some from Europe via its blockade-running ships. The next most reliable source was entire medical kits that were captured whenever a Union field hospital was overrun. Also, in Memphis the South frequently was able to trade tobacco and coffee for medical supplies shipped from the North.

Postoperative infections posed a grave problem for recuperating soldiers. Thick pus and bloody pus originally were thought to be symptoms of healing and recovery. Gangrene, however, was known to be dead or dying tissue, and if not removed it could lead to death. With the patient anesthesized a surgeon would cut away dead flesh and inject bromine into wound margins. Open wounds then were packed with a bromine-drenched dressing. Gangrene patients were housed in a ward apart from all other sick and wounded men.

> Like the dim caverns of the catacombs where, instead of the dead in their final rest, there were wasted figures burning with fever and raving from the agony of splintered bones, tossing restlessly from side to side with every ill, it seemed, which human flesh was heir to. From the rafters the flickering oil lamp swung mournfully, casting a ghastly light upon the scene beneath.
>
> *[Pvt. Alexander Hunter, 17th Virginia Infantry Regiment, writing about a night in a field hospital near Petersburg, Virginia]*

Causes of Disease: Disease posed the deadliest threat to a soldier. When the war first began surgeons were generally ignorant about factors that caused disease and infection. Significance of sanitation and sterility were not understood. Camp hygiene was neglected, and soldiers' food was atrocious.

> The diet of the Civil War soldier was somewhere between barely palatable, and absolutely awful.
> *[Department of History, Ohio State University]*

A soldier's diet included pickled beef when it was available, but it often was spoiled. Pork, if available, was salt-cured, mostly fat, and usually rancid. No fruits or vegetables were provided, although soldiers sometimes could get them from local farmers. Dietary staples were hardtack and cornbread. Hardtack is a type of biscuit or cracker made from flour, water, and salt – nothing more. Although not nutritious, it is non-perishable and long had been a staple on sea voyages of the era. As the name implies, hardtack is *hard* and must be soaked in water or coffee before it can be eaten. Coffee and tobacco were mealtime basics. Malnutrition was common. The most frequent illness among soldiers was gastrointestinal distress.

> Fresh tender meat is very scarce and becoming scarcer every day. Butter, eggs, milk, wines, and brandies are to be obtained only in small quantities.
> *[Richmond Whig newspaper, 26 May 1864]*

A host of factors contributed to stomach ailments. Latrines often drained into streams and springs, contaminating the drinking water source. Military food was poorly treated and stored, and it lacked Vitamin C that could prevent scurvy. Most available medications were botanicals, which were of limited medicinal value. Chronic diarrhea and dysentery, with their debilitating discharge of mucus and blood, caused more deaths than battle wounds.

In the camps young men were billeted together with hundreds of strangers. Bad drinking water, harsh winter weather, poor unheated shelter, and filthy living conditions took a lethal toll. In southern states in the summer, typhoid fever (aka, *camp fever*), yellow fever, and malaria ran rampant. There were no antibiotics, so surgeons prescribed coffee, quinine, and whiskey. Outbreaks of measles became common. There were about 1,200 soldiers in the 12th North Carolina Regiment, and 820 of them contracted measles during a five month period in 1861. During winter months with no heat and little shelter, the common cold often degenerated into pneumonia, an inflammation of the lungs. A medical inspection team visited one

C.S. Army camp and later reported:

> The camp was littered with refuse, inedible food, and rubbish, all in an offensive state of decay, with *[feces]* in pits and also lying atop the ground. Mosquitoes and flies filled the air, and rats scurried about.

By mid-1863 camp sanitation improvements served to lower the disease rate. In a camp environment in the summer, soldiers were required to bathe two times each week and change their clothing once each week. Trash and refuse had to be burned or buried. Latrines were relocated to sites far from drinking water sources. Even in freezing weather and during rainstorms, soldiers were required to use latrines, not merely the nearest convenient area.

Surgeons learned to use available medicine more effectively. Chronic dysentery and diarrhea, the most lethal killers, were treated with opium or with a mixture of chalk and mercury. Pneumonia and bronchitis were treated with quinine, mustard plasters, and sometimes opium. Quinine became the most effective cure for malaria, but in its absence turpentine was administered. When practical, camps were set up far from swamps and stagnant water, and this reduced the number of mosquitoes and the diseases they carried. Lessons learned during the war would speed the advancement of medical knowledge in decades to come.

Newly Constructed General Hospitals: New military hospitals in Richmond cared for wounded and sick soldiers who lived long enough to reach them via rail and wagon. After the war began five enormous medical complexes in the city were built specifically as Confederate States military hospitals:

> Chimborazo General Hospital: This hospital was named for the high plateau on which it was built, which in turn had been named for Mount Chimborazo in Equador. Work on this vast complex overlooking the James River began early in 1861. When it opened that fall on 17 October it was the largest military hospital complex in the world. Its 120 buildings included bakeries, a soap factory, ice houses, kitchens, blacksmith shops, apothecary shops, etc. There were over 90 separate wards for the sick and wounded. Each ward had a

shingle roof, wood plank floors, and whitewashed walls. Wards were warmed by a wood stove during cold months and lighted by a single candle at night. Large natural springs provided a reliable source of fresh water.

Total hospital capacity was 3,000 beds. The hospital had its own farm for row crops and for raising goats and cows. A private canal trading boat enhanced the staff's ability to obtain supplies and food from the Richmond area. In effect, the hospital functioned like a small city. With huge nursing and support staffs, Chimborazo became one of the world's most modern and efficient medical centers of the era.

Wards initially were grouped into five separate divisions so that soldiers from Virginia, Kentucky, Missouri, Maryland, and Tennessee could be in the same division with other men from their state. Although well intended, this segregation by state lasted only about one year. During the war the hospital cared for roughly 76,000 wounded and sick soldiers. The mortality rate ranged between eight and ten percent, an exceptionally low death rate for military hospitals of the day. Surgeon-in-chief: Dr. James B. McCaw.

After the war ended in April 1865 the hospital initially housed wounded soldiers from both the Confederacy and the Union, but by mid-summer all patients had left. Chimborazo then became home for the new federal Bureau of Refugees, freedmen who had no place to live and nothing to eat. These refugees became the "poorest of the poor" in America.

Howard's Grove General Hospital: A rural area known as Howard's Grove had been a popular picnic and recreation area on the outskirts of Richmond before the war. Before hostilities began in 1861 the Confederacy converted the area into a military training ground for Confederate troops that flocked to Richmond. After the initial land battle at Bull Run (aka, *Manassas*) in July 1861 the Confederacy began building a huge hospital complex at Howard's Grove. The facility began accepting patients nine months later, but construction continued. By the end of 1862 the hospital had 659 beds, and capacity soon expanded to over 1,800 beds.

> In the vast wards of Chimborazo, Winder, Jackson, and Howard's Grove *[hospitals]* we have ample space for all present and prospective wounded. We have reason to believe that the organization of these great hospitals is as nearly perfect as the straitened condition of the Confederacy will permit.
> *[Richmond Whig newspaper, 26 May 1864]*

The hospital complex included bakeries, warehouses, laundries, recreational facilities, and of course the necessary "dead houses." Part of the hospital was reserved for smallpox cases. Surgeons-in-chief: Drs. James Bolton, T.P. Temple, and P.M. Palmer.

Women played a key role at Howard's Grove and at other large Confederate military hospitals. They read letters and Biblical passages to sick and wounded men, wrote letters for them, held the hands of the dying, served as practical nurses and matrons, and helped to feed the disabled. To hospitalized soldiers such women were angels in disguise. After the war the hospital reverted to civil control and continued in use as the Union Smallpox Hospital.

Jackson General Hospital: Named for "Stonewall" Jackson, this large complex opened its doors on 29 June 1863 with a 1,980-bed capacity. Later it was enlarged so that it could accommodate 2,500 patients. Situated in the western suburbs of Richmond, the hospital featured a vast 40-acre vegetable garden. Initially the hospital had four divisions so that soldiers from a given state could convalesce together. One report in 1863 notes:

– Division 1 – South Carolina soldiers – 380
– Division 2 – Georgia soldiers – 450
– Division 3 – Georgia soldiers – 551
– Division 4 – Louisiana soldiers – 600

On 9 May 1864 a report noted that the hospital was "filled with North Carolina, South Carolina, and Georgia troops" from battles in northern Virginia. In March 1865 as the war neared its end, free Negro hospital workers made up three companies of Confederate troops who served "in the lines under fire." After the war ended a part of the hospital was converted for use as a Union prison.

Stuart General Hospital: Before the war began the Virginia state fairgrounds had been located on land called Monroe Park. Early during the war the old fairgrounds were used as a drill and bivouac area for Confederate troops based in the Richmond area. Later the City Guard barracks were built on this property.

Late in the war the barracks were converted to a hospital complex that opened on 2 June 1864. Named for Gen. "Jeb" Stuart, the hospital consisted of 16 primary buildings with necessary kitchens, bakeries, ice houses, etc. The wards could accommodate about 500 patients. Surgeons-in-chief: Drs. S. Meredith and R.A. Lewis.

Richmond never experienced a direct Union attack, so Stuart and other large Richmond military hospitals functioned in safety until April 1865. After the Confederate government evacuated Richmond the hospital became Camp Winthrop, a post hospital for Union troops until 31 December 1865.

Winder General Hospital: In terms of sheer patient capacity, Winder General Hospital became the largest hospital in the Confederacy. Named for Gen. John Winder, the facility first opened in 1862 with beds for 3,000 soldiers. Within a few months this capacity grew to 4,300 beds, including a tent division for overflow usage. The hospital had 98 buildings including kitchens, surgical wards, bathhouses, bakeries, etc.

> In the middle of Camp Winder streets were water closets with barrels underneath. These barrels were carted off to the city daily, and the contents were used to make salt-peter, which was used to make gunpowder.
> *[memoir of William Curtis, CSA, 2nd North Carolina Cavalry, circa 1880]*

Including the tent division, Winder Hospital had seven divisions for wounded and sick soldiers. Deep wells and natural springs were within the huge complex, which included 125 acres of farmland for row crops and livestock. This enormous medical facility had its own fire brigade. Surgeon-in-chief: Dr. Alexander G. Lane.

Late in the war in concert with nearby Jackson Hospital, civilian employees and recuperating soldiers from Winder Hospital formed an infantry battalion. They fought at the Battle of North Anna in May 1864 and in struggles during the desperate spring of 1865. Two months before the end of the war the two hospitals formed a company of volunteer free Negro soldiers, and they manned defensive fortifications around Richmond. After the war ended and the Union Army occupied the city, the hospital was renamed Camp Grant. It was used as the federal 24th Army Corps headquarters.

Smaller General Hospitals: At the start of the war a variety of hotels, schools, tobacco warehouses, etc. were pressed into use as hospitals to care for wounded and sick soldiers in Richmond. Many of these smaller hospitals were used only for a short time, or used off and on after major battles. Others remained in use throughout the war. The 28 largest of these general hospitals were assigned numbers by the C.S. Army. They are listed below:

General Hospital No. 1 (aka, *City Home Hospital* or *Alms House Hospital*): This structure was built as a Poor House by the city, but in June 1861 it was rented to the Confederacy as a military hospital. However, it initially was used to house Union prisoners after the Battle of First Bull Run. After that influx it served as a hospital for Confederate soldiers until December 1864. During that time it had a large 500-patient capacity. Surgeon-in-charge: Dr. Charles B. Gibson.

The hospital suffered heavy exterior damage when the nearby Confederate gunpowder magazines detonated and burned on evacuation night. After the war ended the facility reverted to city use and was rented to Virginia Military Institute. Later it again became the city Poor House, and eventually it evolved into Richmond Nursing Home.

General Hospital No. 2 (aka, *Bailey's Factory Hospital*): This three-story A-roofed building had been the tobacco factory, or tobacco warehouse, of Bailey & Company. It was used as a military hospital in 1861 and 1862, and little else is known about it. Surgeon-in-charge: Dr. James M. Holloway.

General Hospital No. 3 (aka, *Byrd Island Hospital* or *Gilliam's Factory Hospital*): This building had been the tobacco factory of Gilliam Brothers Company. It began use as a Confederate military hospital on 15 August 1861, three weeks after the Battle of First Bull Run. The hospital had beds for 225 patients. Surgeon-in-charge: Dr. A.Y.P. Garnett.

> *[That day]* in the General Hospital we had no dinner, and supper consisted of a piece of hard, dry, stale, light bread and coffee. I did not care much whether I lived or died. *[memoir of William Curtis, CSA, 2nd North Carolina Cavalry, circa 1880]*

General Hospital No. 4 (aka, *Baptist Institute Hospital*): Originally the Richmond Female Institute, a Baptist academy for girls, the building opened as a military hospital in the summer of 1861. It was still used as such through most of 1864. It had a capacity of over 300 beds, but after early 1863 it housed wounded and sick officers only. Surgeon-in-charge: Dr. James P. Read.

After the war the building was used by Union troops as administrative office space. In later years it would become part of the University of Richmond.

General Hospital No. 5 (aka, *Kent Hospital*): This building was the former home of Kent, Paine & Company, a dry goods department store. It opened as a military hospital sometime before June 1862, when it had 97 patients. The structure was destroyed by fire during evacuation night in Richmond.

General Hospital No. 6 (aka, *Keen Hospital*): This facility initially was the home of Keen, Baldwin & Company, a dry goods store. Shortly after it opened as a hospital in June 1862 it had 86 patients. The building burned to the ground during the night when Richmond was evacuated.

General Hospital No. 7 (aka, *Bacon & Baskerville Hospital* or *Receiving and Wayside Hospital*): This structure originally housed the firm of Bacon & Baskerville, wholesale grocers and commission merchants. Little is known about its role as a military hospital. It was destroyed in the fire that swept through much of Richmond on evacuation night.

General Hospital No. 8 (aka, *Saint Charles Hospital*): Built in 1846 on the site of a former tavern, the large structure opened as the Saint Charles Hotel. It became a military hospital in July 1861, and a report dated the next month on 13 August states that 460 patients were then being treated there. The hospital closed its doors in late 1863. Superintendent: H.A. Dudley.

General Hospital No. 9 (aka, *Seabrook Hospital* or *Receiving and Wayside Hospital* or *Receiving and Distributing Hospital*): This huge building had been built by John Seabrook in 1810 as a warehouse. One of the first buildings in Richmond to be converted into a military hospital, it opened in early 1862 with a capacity of 900 patients. It functioned as a primary receiving hospital because it was near the Virginia Central Railroad depot. Surgeon-in-charge: Dr. John J. Gravatt.

> *[In the hospital]* I had little to eat in three days. I came to the conclusion that evening that I had to do one of two things – either submit and die, or determine to get well. *[memoir of William Curtis, CSA, 2nd North Carolina Cavalry, circa 1880]*

This hospital remained in use until the end of the war. A report dated October 1864 shows that during the preceding three month period 10,107 patients were admitted. Of those, 9,663 were transferred to other hospitals, 96 died, and 348 remained at Seabrook.

General Hospital No. 10 (aka, *Union Hotel Hospital* or *United States Hotel Hospital*): This elaborate building was built in 1817 as the Union Hotel, an ultra-modern facility for its day. In 1838 it housed the Medical College of Virginia, and seven years later it reopened as the United States Hotel.

The Confederacy used the rambling building as a military hospital from July 1862 until the end of the war in 1865. It could house over 300 patients and generally was reserved for officers only. Surgeon-in-charge: Dr. William A. Carrington.

The structure survived the war and became the home of Richmond Theological Seminary, later to become known as Virginia Union University. Still later the building housed the Richmond Methodist Mission. After standing vacant for several years the building was demolished in 1911.

General Hospital No. 11 (aka, *Globe Hospital* or *Florida Hospital*): Built in 1853, this building originally was the home of a tobacco factory owned by James H. Grant & Company. When the Confederacy turned the building into a hospital in June 1862 it had room for about 150 patients. This relatively small hospital remained in use for slightly over a year. It closed in September 1863.

> *[Soldiers]* who overcame their illness during the war owed their recoveries less to the ingenuity of contemporary medicine than to grit and chance.
> *[Ina Dixon, Civil War Medicine, 2013]*

General Hospital No. 12 (aka, *Banner Hospital* or *Wayside Hospital* or *Grant Hospital*): Constructed in 1853, this brick building originally was a combination tobacco factory and warehouse owned by William H. Grant. On 2 December 1861 it opened as a Confederate military hospital with over 250 beds. On 6 August 1863 it was designated as a Wayside Hospital for soldiers on furlough and those who had been honorably discharged. After the war the federal occupying troops used the building as a barracks.

General Hospital No. 13 (aka, *Castle Thunder Hospital* or *Prison Hospital* or *Lunatic Hospital*): This building formerly had been the tobacco factory of Christian & Lea Company. The brick four-story flat-roof structure opened as a military hospital sometime prior to June 1862. At that time military mental patients had been housed in a facility known as Castle Thunder. The mental patients soon were moved to General

Hospital No. 13, but they were kept in a part of the building apart from physically sick and wounded soldiers. Surgeon-in-charge: Dr. W.W. Coggin.

In addition to the mental cases, imprisoned soldiers were confined in yet another separate part of the hospital. Total capacity was about 150. One report notes that the hospital was "clean and neat to the last degree." After the war the building was used by occupying federal troops.

General Hospital No. 14 (aka, *Second Georgia Hospital*): Not much is known about this Confederate hospital. It opened sometime before mid-1862, and in June of that year it had 65 patients, although capacity reportedly was in excess of 125.

General Hospital No. 15 (aka, *Crew's Hospital* or *Crew & Pemberton Hospital* or *Conrad & Crew Hospital*): This four-story brick building had been the tobacco warehouse for Crew & Pemberton. It opened as a Confederate military hospital in June 1862 with a capacity of 200 patients. Three months later the hospital was converted into a barracks for soldiers who were traveling or convalescing. Surgeons-in-charge: Drs. H.C. Scott, G.B. Moore, and W.T. Bell.

General Hospital No. 16 (aka, *First Georgia Hospital*): Before becoming a military hospital this facility had been a tobacco factory for John E. Whitlock & Company. The date when it first became a hospital is not known, but a report dated June 1862 shows that the facility then had 66 patients. Capacity was stated to be "over 130." The length of time that this hospital was in operation is not known. The building was destroyed by fire during evacuation night.

General Hospital No. 17 (aka, *Fourth Georgia Hospital*): Before it was used as a medical facility this building had been a tobacco warehouse for the firm of Smith & McCurdey. The building opened as a Confederate military hospital on 19 May 1862. The next month a report showed that the hospital had 86 patients, and capacity was stated to be "over 125." The building burned to the ground during evacuation night.

General Hospital No. 18 (aka, *Greanor's Factory Hospital*): This structure was built in 1853 as a tobacco factory for John H. Greanor, a member of the city council and a soldier in the C.S. Army. The facility opened as a hospital in early 1861. On 25 September of that year a report showed that it housed 260 patients. Total capacity is not known.

After renovations, two years later on 2 June 1864 the facility reopened as a shoe factory. About 300 workers toiled on shifts, 24 hours per day, to provide shoes and boots for Confederate troops.

General Hospital No. 19 (aka, *Third Georgia Hospital*): Prior to use as a military hospital this four-story facility had been the tobacco factory for A.W. Taylor Company. This small hospital opened in June 1862 with a capacity "over 75." It closed 18 months later in December 1863.

General Hospital No. 20 (aka, *Royster's Hospital* or *First Alabama Hospital*): This facility formerly was the tobacco factory of two brothers, J.B. Royster and A.L. Royster. It opened as a Confederate military hospital in early 1862, and by June of that year it had 44 patients. Total capacity is not now known but was reported as "very large." The date when the facility ceased to be used as a hospital is not known.

General Hospital No. 21 (aka, *Gwathmey Factory Hospital* or *Maryland Hospital* or *Prison Hospital*): This large structure originally was a tobacco factory for R.A. Mayo Company, and later for the Gwathmey Tobacco Company. The six-story brick facility was one of the tallest buildings in Richmond. It opened as a Confederate military hospital in early 1862 with a capacity "over 700." In December of that year it was the primary hospital for smallpox patients during the smallpox epidemic. Later most smallpox patients were moved to the larger Howard's Grove Hospital.

In the summer of 1864 the hospital was used to house and treat about 650 wounded Union prisoners. Incidentally, in November 1864 the Union prisoners who were physically able to participate were allowed to hold a *straw vote* for U.S.

President (their erstwhile commander-in-chief). Abraham Lincoln got 154 votes, but Gen. McClellan garnered 302 votes. Hospital surgeon-in-charge: Dr. G.W. Semple.

General Hospital No. 22 (aka, *Howard's Factory Hospital* or *North Carolina Hospital*): The three-story brick building was originally a tobacco factory for George D. Howard Company. It opened as a military hospital in 1862 and closed two years later. Capacity was said to be about 110 patients. Surgeon-in-charge: Dr. Addison C. Fox, who was praised for his success with "desperate cases" in May 1863.

General Hospital No. 23 (aka, *Ligon Factory Hospital*): This facility opened as a Confederate military hospital on 31 May 1862. Prior to that time the building had been a tobacco factory owned by John L. Ligon. Hospital capacity is not known, but it was reported that 45 employees worked there. The exact closing date has been lost in the fog of history, but the hospital was still in operation in late 1864.

General Hospital No. 24 (aka, *Harwood's Hospital* or *North Carolina Hospital*): This three-story brick building had been the tobacco factory for George D. Harwood. It opened as a hospital in the summer of 1861, and in July that year it took in several hundred Union prisoners who had been wounded and captured at the Battle of First Bull Run. Capacity was said to be "over 225."

After July 1864 the hospital was used almost exclusively for Confederate soldiers from North Carolina. Surgeon-in-charge: Dr. O.F. Manson. Matron: Lee Sanxay.

General Hospital No. 25 (aka, *Randolph's Hospital* or *Texas and Arkansas Hospital*): Not much is now known about this military hospital. It opened in 1862 and was still in operation in March 1863. Capacity was reported to be about 225. At times wounded Union prisoners were hospitalized there.

General Hospital No. 26 (aka, *Springfield Hall Hospital* or *Masonic Hall Hospital*): This fraternal meeting hall was owned by the Springfield Division, Sons of Temperance. It also was used by two Masonic lodges. Built in 1850, the two-story brick building opened as a small military hospital in mid-1861. At that time it was jointly operated by the (1) Soldier's Aid Society, (2) Trinity Methodist Church, and (3) Leigh Street Baptist Church. Capacity in this little hospital was about 40. Surgeon-in-charge: Dr. John Know.

General Hospital No. 27 (aka, *Gangrene Hospital* or *Company G Hall Hospital*): Originally a social hall and militia drill hall for Militia Company G, this small building was converted into a military hospital in March 1862. Capacity was about 50. Many Confederate gangrene casualties were treated here. The hospital closed in July 1863.

General Hospital No. 28 (aka, *Danville Railroad Hospital* or *Railroad Shops Hospital*): A huge complex of maintenance shops of the Richmond & Danville Railroad were converted into a Confederate military hospital, and it opened on 10 July 1862. The single-story buildings were of brick construction. The hospital served primarily as a surgical unit, and capacity was reported as 416 beds. Several nearby natural springs provided an abundance of fresh water. Surgeon-in-charge: Dr. A.S. Miller.

> Achilles: But even for me, I tell you, death and the strong force of fate are waiting. There will come a dawn or a sunset or a high noon when a man will take my life in battle too – flinging a spear perhaps, or whipping a deadly arrow off his bow.
> *[Homer, The Iliad, circa 760-710 BC]*

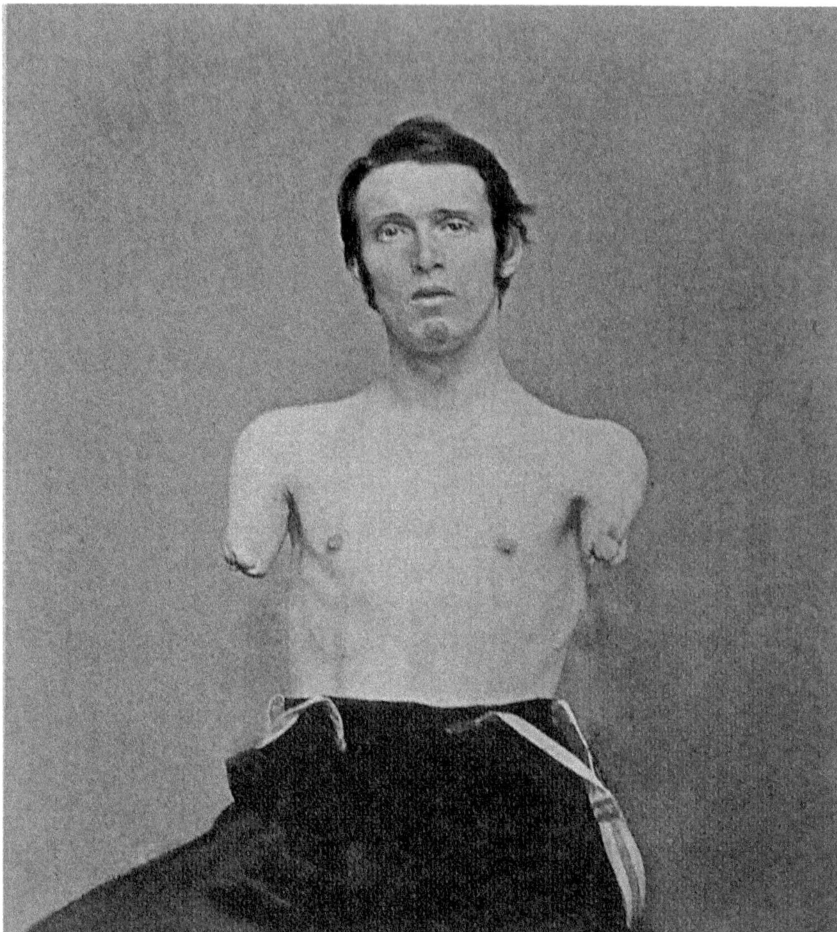

Alfred A. Stratton, a private in the 147th New York Infantry, suffered the loss of both arms in battle. His arms were "carried away by a solid cannon shot" fired from Confederate defenses at Petersburg, Virginia, on 18 June 1864. After recovery, Stratton was discharged from the U.S. Army. He became a husband and a church pastor, but he died at age 29 due to complications from his war injury .

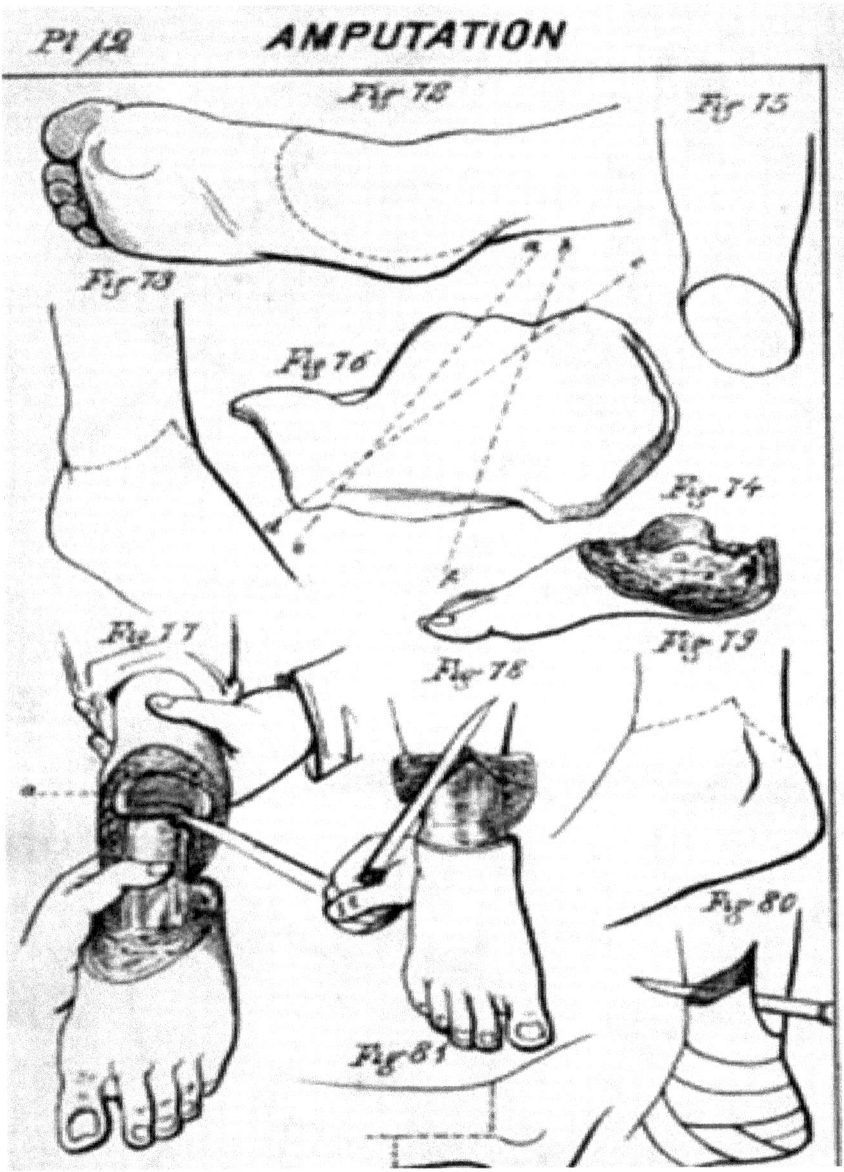

Page from a graphic military surgery manual used by Confederate and Union armies during the war. Most new surgeons in both armies had never treated a gunshot wound, and very few had attempted an amputation. When amputating a leg, arm, or hand the limb would be severed above the wound and the shattered portion of the bone. Skin would be removed from the severed limb and stitched to the stump.

A portion of a flyer advertising Left-Handed Penmanship by soldiers or sailors who had lost their right arm during the war. The price of admission was 25 cents. Money received would be used to support disabled amputees.

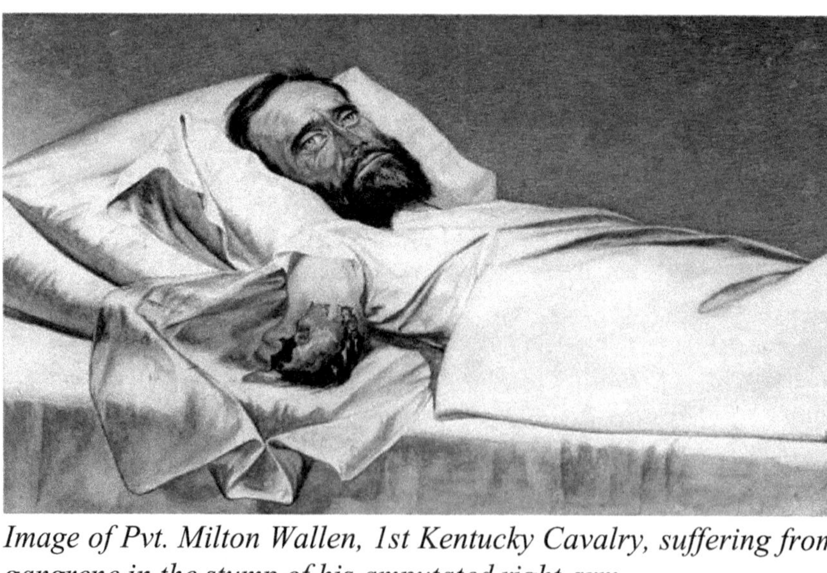

Image of Pvt. Milton Wallen, 1st Kentucky Cavalry, suffering from gangrene in the stump of his amputated right arm.

Display of a prosthetic leg from the 1860s era. One major supplier to amputees from both armies offered his "Anatomical Patent Leg" for $150 and his economical "Army and Navy Leg" for $75.00.

Nurses, soldiers, and others pose at a make-shift field hospital near Fredericksburg, Virginia, in May 1864.

Prisoners of War

> Section 1: . . . All prisoners of war taken, whether on land or at sea, during the pending hostilities with the United States shall be transferred . . . to the Department of War *[which]* shall provide for the safe custody and sustenance of prisoners of war, and the rations furnished to prisoners *[of war]* shall be the same in quantity and quality as those furnished to enlisted men in the army of the Confederacy. *[An Act Relative to Prisoners of War approved by the C.S. Congress, 21 May 1861]*

Prisoner Parole and Exchange: In warfare military losses are not confined to combatants killed in battle, wounded, or missing. Many soldiers are captured and will become prisoners of war (POW). In antiquity the weak, wounded, and infirm captives were of little use to the victors and thus were routinely killed. Healthy captives were retained as chattel slaves.

Customs had changed by the era when the U.S. Army and C.S. Army clashed in the early 1860s. During four years of warfare both armies captured many thousands of enemy combatants. Years later in 1901 the U.S. Record and Pension Office would try to document soldiers who were taken captive during the war. Their best estimate is that 211,000 Union troops were captured, and roughly 198,000 Confederate soldiers met the same fate.

During early months of the war neither side was prepared to house, feed, clothe, and guard large numbers of military prisoners. Therefore both sides practiced the traditional European system of *parole and exchange*. A prisoner would be paroled if he signed an agreement not to fight again until he was exchanged for a prisoner held by the other side. Pending parole and exchange, POWs were briefly confined, fed, and guarded by their captors.

Throughout 1861 the Union government avoided participation in the exchange system. President Lincoln feared that participation might be considered as recognition of the Confederate government, which he wanted to avoid. Nonetheless, Union and Confederate military commanders unofficially exchanged POWs on humanitarian grounds. At that time the South had captured many more prisoners than the North. Unable to exchange excess POWs on a one-for-one basis, southern commanders were obliged to retain, feed, and guard them. In general these prisoners were well treated and were allowed to write letters to their families.

Letters from Union POWs who had not been paroled reached northern newspapers, and citizen petitions forced the U.S. Congress to act. On 11 December 1861 the Congress passed a joint resolution demanding that Lincoln "inaugurate systematic measures for the exchange of prisoners in the present rebellion." Lincoln relented, so Union and Confederate officials met on 23 February 1862 to reach a formal agreement.

The so-called Dix-Hill cartel established exchange equivalency. For example, privates would be exchanged on a one-for-one basis, and a colonel would be exchanged for 15 privates. Each government would appoint an agent to monitor the process. Direct exchanges between opposing commanders also were authorized. Further, non-combatants (disloyal citizens, civilian employees of the military, etc.) could be exchanged. Significantly, if one side held excess POWs they would have to be paroled within 10 days of their capture date under the following condition:

> Such parolees are prohibited from returning to the military in any capacity, including the performance of field, garrison, police, guard, or constabulary duty.

Failure of the Exchange System: The POW exchange system began to fail in late 1863, and by 1864 it was abandoned altogether. The South was feeling a military manpower pinch. The C.S. Army had an increasingly limited pool of available recruits. Each battle casualty and each soldier captured became harder and harder to replace, but the populous North had no such problem. Aware of the manpower shortage in the South, the Union stopped exchanging prisoners. Gen. Ulysses Grant, a Union commander, wrote that retaining captured Confederate soldiers in prison camps gave the Union a "decided military advantage."

The South had its own reason to abandon the exchange system. On some occasions when opposing commanders tried to exchange prisoners, the Union wanted to exchange a runaway slave in lieu of a captured Confederate soldier. That was unacceptable to the South. It would have given the Union another soldier to fight and would have given the South no such advantage.

When the exchange system fell apart the number of POWs skyrocketed on both sides. Both North and South began converting old vacant barracks complexes into prison camps. Also new camps were hastily thrown together. In total at least 154 such military POW camps would spring up during the war.

Inhuman overcrowding became the norm on both sides. Filthy squalid POW camps, little food, absence of medical care, exposure, and disease combined to make life in such camps horrible beyond words. Mortality rates on both sides were high. About 56,000 soldiers died in POW camps. The two most notorious prisons were (1) Elmira Prison Camp, a Union POW camp in western New York state, and (2) Andersonville Prison Camp (aka, *Camp Sumter*), a Confederate POW camp in rural Georgia. At Elmira, 26 percent of the Confederate prisoners died in captivity. At Andersonville, 28 percent of the Union prisoners died.

POW camps in the deep South were plagued by high disease rates due to the hot and humid summer climate. Prisoners in northern camps were faced with bitter below freezing cold in winter. Medical attention and medicine were minimal or totally lacking. Food was scarce and often inedible. Prisoners starved to death, froze to death, and died of disease. Acute dysentery was the biggest killer, followed by typhoid fever, scurvy, starvation, and exposure to the elements.

> As we entered the place a spectacle met our eyes that almost froze our blood with horror. . . . Before us were *[men]* that had once been active and erect – stalwart men, but now *[they had become]* nothing but walking skeletons covered with filth and vermin. . . . Can this be Hell?
> *[Robert H. Kellog, USA, writing about his arrival at Andersonville POW camp in 1864]*

Life in the Camps: Prisoners who were nearing death frequently became catatonic, and thereafter nothing seemed to reverse their physical and mental decline. Some took refuge in suicide by crossing the "dead line" fence, knowing that guards would shoot them. Even battle hardened veterans found themselves unprepared for horrors of life in a POW camp. Prior to capture they had written letters that extolled the carnage of warfare. However, they found their POW experience to be far worse. One historian later would write:

> A Union propaganda campaign following the war decried the dire conditions of Confederate prisons, while ignoring their own.

Violence periodically flared among inmates, At Andersonville prison in Georgia a group of Union prisoners formed the "Raiders." Desperate for food and shelter, they banded together and terrorized other inmates, stealing food, blankets, and whatever they wished. Finally other inmates responded. They formed the "Regulators" and took justice into their own hands. They overpowered the Raiders and restrained them. Regulators then pled their case to Capt. Henry Wirz, the Confederate camp commander. Wirz told the Regulators to form a jury and try the offenders. He agreed to sanction whatever punishment the jury decided upon. The prisoner jury sentenced the six Raiders to death, and Wirz ordered a scaffold to be built. With his blessing the jury hanged all the Raiders on 11 July 1864.

Letters and diary entries do much to illustrate misery in POW camps, both in the North and in the South:

> The camp was covered with vermin all over. You could not sit down anywhere. You might . . . pick all the lice off of you, and sit down for a half a moment, and when you got up you would be covered with them. Where the *[feces]* was emptied it was all alive. There was a regular buzz there all the time, and it was covered with white maggots.

> *[I]* drew a half of a pint of moldy rice and a small piece of pork for a day's ration.

> Three rows of bunks, one above the other, were built on the sides of the *[barracks, and we slept on]* the bare rough planks. Shavings, straw, or bedding of any kind was not permitted, as the authorities said they produced vermin.

> A poor crippled *[prisoner]* was shot for stepping inside the dead line. He said he was so miserable he wished to die, and *[he]* took this means of having his wish gratified.

The small rations, of such poor quality, *[are]* killing the men off at a terrible rate. There are now over one hundred bodies at the gate to be carried to the "dead house."

[We] drew "fresh meat" that the rebel quartermaster calls beef, but he can't fool old soldiers with his mule and horse flesh. It might have been pretty good had they brought it in within a week after its death, or had given us a large enough piece to allow for the maggots.

The pants I had . . . were in such a bad condition that I wore nothing but my underwear. . . . *[If]* we had been out in the world in such a costume, one might have mistaken us for scarecrows.

Many more prisoners from Grant's and Sherman's armies *[arrived in camp]*, among them a number of one-hundred-day men whose term of service had almost expired.

I wrapped my feet in old rags . . . but in the winter we were compelled to stand in the snow every morning for roll call. My feet and shins were badly frozen, *[and]* they had the appearance of gobbler's legs.

I have witnessed many horrible sights, but none to compare with what I saw today – a man lying on the bank of the *[latrine]* being eaten to death by maggots. They could be seen issuing from his eyes and mouth, and his body was completely raw . . . . We left him alone to die a miserable death.

He bore his prison life bravely but has at last fallen victim to ill treatment and starvation. He was loved by all who knew him.

A recruit of our company, and a bunk mate of mine, died today, *[and]* I helped carry his body to the dead house. I passed through the hospital *[tent]* on the way back. . . . A man is never admitted to the hospital *[tent]* until there is no hope for his recovery.

Every day brings the sad news of the death of one or more

of our comrades. Death! Nothing but death! Rations small. Almost starved.

The poor fellows died rapidly, despondent, homesick, hungry, and wretched. I have stood day after day watching the wagons carrying the dead outside to be buried. . . . The name of the company and regiment of the dead *[prisoners]* were written on a piece of paper and put into a tightly corked bottle, and buried with the corpse.

[I] have had nothing to eat for two days. *[I]* am so weak I can scarcely walk, and hungry enough to eat anything.

Blankets were brought into camp to be distributed among men who were entirely destitute of shelter. Blankets were drawn by lot, but before drawing the following oath was taken: "You do solemnly swear that you are entirely without shelter, that you do not possess a blanket, or do not bunk with anyone who does."

Rations are smaller and of poorer quality. Wood more scarce. Lice plenty. Shelter worn out. Cold weather coming on. I am getting too weak to move about. . . . I have scurvy of the gums.

We have a little wood *[to use to build a fire]*, but it is so wet we cannot *[start a fire to]* cook with.

Rained all night. . . . We have not had a bite to eat for seventy hours. I am almost famished.

In the latter part of the winter, crackers were substituted in place of soft bread.

Weakness and starvation caused me to lose my sight. We are all so weak from starvation that we have been obliged to suspend work on our shanty.

It rained all night. Could not sleep on account of the cold and lice. It seems as if for every louse we burned, two more came in its place.

By December, 1,600 half naked men, entirely destitute of

> blankets, stood in ankle-deep snow to answer the morning roll call.

Prisoners coped as well as they could. POWs formed a debate society at Johnson's Island prison in the North. In many camps, inmates congregated in ethnic groups from which a variety of languages could be heard: German, French, Italian, etc. In the absence of money, barter systems allowed prisoners to trade what few valuables they had. However, the primary activity was talk of escape. The tantalizing wistful ever-present dream of escape dominated idle conversation.

Escape: There were many escape attempts, and a surprising number were successful. Some were relatively simple. A few skeletal POWs feigned death so well that they were carried outside the stockade and were left to be buried later. When the opportunity presented itself they simply walked away. At Camp Douglas in Illinois a few prisoners smeared their sunbaked faces and hands with charcoal. Then they blended in with the guards' Negro servants and calmly walked out of camp at the end of the workday. As farfetched as that scheme may seem, it often succeeded. It became so successful that the camp superintendent outlawed use of Negro labor.

Digging tunnels was the most popular method of escape. The best known successful tunnel escape would take place at Libby Prison Camp near Richmond. Using nothing more than clamshells and case knives, Union prisoners would dig a 60-foot tunnel. Over a period of months they concealed excavated dirt in their haversacks and scattered it over the camp grounds. In the dark of night 109 Union POWs crawled to freedom through what became known as the Great Yankee Tunnel.

Four Typical POW Camps: In the North and in the South, prison camps for captured soldiers sprang up during the war. When the parole and exchange system was abandoned, POWs by the tens-of-thousands were sent to such camps. Four camps are highlighted below. The first two are Confederate POW camps. The last two are Union camps:

Salisbury Prison Camp (Confederate): Near Salisbury, North

Carolina, this camp opened in the summer of 1861 after the Confederate victory at the Battle of First Bull Run (aka, *Battle of First Manassas*). About 120 captured Union soldiers were shipped by rail to Salisbury and confined in a former cotton mill. Initially conditions in the new prison were adequate. Sanitation was acceptable, and rations were satisfactory. The masonry structure protected POWs from the elements, and the mortality rate hovered around two percent. Only one inmate in each fifty died.

However, collapse of the parole and exchange system changed everything. Prison population zoomed to over 9,000 POWs. Overcrowding, disease, lack of food, filth, and absence of medical care wreaked havoc on the inmates. The mortality rate soared up to 28 percent.

Belle Isle Prison Camp (Confederate): This prison opened in 1862 early during the war. It was located on a 55-acre island in the James River near Richmond. No barracks were built, and POWs were lodged in thin canvas tents. Over 30,000 prisoners eventually filled the prison compound. Lucius Chittendon, the U.S. Treasurer during the war, described the Belle Isle prisoners who were released and repatriated after the war ended in April 1865:

> *[They were]* in a semi-state of nudity . . . laboring under such diseases as chronic diarrhea, scurvy, frost bite, and general debility caused by starvation, neglect, and exposure. Many of them had partially lost their *[memory]*, forgetting . . . everything connected with their antecedent history. They resembled, in many respects, patients laboring under cretinism. They were filthy in the extreme, covered in vermin. Nearly all were emaciated *[and]* had to be cared for like infants.

Point Lookout Prison Camp (Union): Located in Maryland, this prison originally was built to hold political prisoners, civilians who actively supported the Confederacy. When the exchange system collapsed late in the war the prison was expanded to hold captured Confederate soldiers. About 4,000

POWs had been housed in the camp in late 1863. By mid-1864 the inmate population had doubled. By the end of the year over 20,000 Confederate POWs were confined there.

The exact mortality rate is unknown, but two sources state that "upwards of 4,000" prisoners died. Overcrowding, poor hygiene, and disease took a heavy toll. One historian, Gary Flavion, would write:

> Human error in overcrowding *[was]* a frequent cause of widespread disease. . . . However, simple error and ignorance devolved into treachery and malicious intent, culminating in tragic losses of human life.

Douglas Prison Camp (Union): Years later this Union prison camp in Illinois would become known as the Andersonville of the North. Initially the facility had been a training base for Union troops. Late during the war the Union converted it into a prison camp. By April 1865 over 18,000 Confederate POWs were imprisoned there. Over 4,200 more had died during captivity. Of those, several hundred had simply frozen to death during the bitterly cold Illinois winter of 1864-1865. Conditions in the camp were best described in a report by an inspector from the U.S. Sanitary Commission:

> The amount of standing water, of unpoliced grounds, of foul *[latrines]*, of general disorder, of soil reeking with miasmic accretions, of rotten bones and emptying of camp kettles . . . was enough to drive a *[civilized person]* mad. *[The barracks are so filthy that]* nothing but fire can ever cleanse them.

In terms of high mortality rates and atrocious human conditions, two camps often are cited as the worst of the worst. Andersonville prison, a Confederate POW camp in Georgia, and Elmira prison, a Union POW camp in New York, are detailed below:

Andersonville Prison Camp: By the start of 1864 heavy fighting raged in northeastern Virginia, the site of many Confederate POW camps. Consequently the Confederacy decided to move thousands of Union prisoners farther south to more secure locations. Further, locations in the deep South promised a means of procuring a larger food supply. One new facility, Camp Sumter, was named for the Georgia county in which it was built. Later it would become better known as Andersonville, the name of the nearby town. The prison opened in February 1864 on 16 acres of land. By summertime the compact camp would cover 27 acres.

A 16-foot-high stockade wall constructed of logs surrounded the prison camp. Three feet below the top of the wall a plank walkway allowed guards to walk back and forth. An elevated "pigeon roost" each 90 feet along the walkway was manned by guards. Eighteen feet inside the stockade wall a waist-high fence had been built. This was the "dead line" designed to keep POWs away from the stockade wall. Any prisoner crossing the dead line would be shot by the nearest pigeon roost guard.

In late February 1864 the first train load of prisoners arrived at Andersonville. Every day during March, April, and May about 400 more POWs arrived by rail. By 30 June slightly more than 26,000 prisoners had been jammed into the compact camp. By August the camp held almost 32,000 POWs. Robert Kellogg, a Union prisoner from Massachusetts, later would write:

> In the center of the *[camp]* was a swamp . . . and a part of this marshy place had been used by the prisoners as a *[latrine]*, and excrement covered the ground, the scent arising from which was suffocating.

By 1865 the C.S. Army, civilians in the South, and Andersonville prison struggled to get adequate food. POWs got very little. Their main ration each day was poorly milled corn meal. In order to be edible it needed to be cooked. Although the stockade was surrounded by a forest, little wood was given to prisoners for cooking fires, or for warmth during cold months. The masses of men jammed closely together turned the prison into a small city.

> The noise of thirty thousand men never really subsided. The stockade echoed all day with a clatter and clamor that rose sometimes to a muffled roar; at nightfall it tapered to a sibilant hum, ceasing altogether only for those whose last sun had already set. Prison society continued into the night . . . here and there the murmur of a prayer vigil rose from

> the tents of dying men. By 10 o'clock those chorals usually broke up . . . and gradually the low moans of the sick swelled to fill the darkness with a mournful lullaby. Only in the gray predawn of three or four o'clock did the stockade approach anything that could have been called quiet.
> *[William Marvel, Andersonville: The Last Depot, 2006]*

At Andersonville the primary threat to survival was disease and a poor diet. One culinary staple, cornbread, was full of vermin and poorly cooked, so widespread gastrointestinal illness resulted. Lice were attracted to living tissue, and maggots were drawn to the dead and dying. Scurvy caused gums to bleed and soften, and teeth fell out. Death quickly lost its shock effect.

On 1 January 1865 the prison published and distributed a flyer titled Rules and Regulations of the Prison. One Union prisoner, Lyle Adair, copied the rules in his diary. There were 13 separate rules, and some of them are listed below:

> There will be two daily Roll Calls, one at eight o'clock *[a.m.]* and one at four o'clock *[p.m.]*
>
> No rations will be issued to a division unless all *[POWs]* are present *[or accounted for]* at Roll Call.
>
> The sergeant of a detachment will report all those sick in his detachment and will carry them, after Roll Call at eight o'clock *[a.m.]*, to the receiving hospital *[tent]*.
>
> Prisoners have the privilege to write *[letters]* twice a week. No letter may be over one page in length and must contain nothing but private matters.
>
> No prisoner *[shall]* cross the dead line, nor speak to any sentinel on post, nor attempt to buy or sell anything to a sentinel. Sentinels *[have]* orders to fire on anyone crossing the dead line or attempting to speak to or trade with them.
>
> It is the duty of detachment sergeants to carry men who die in quarters immediately to the receiving hospital *[tent]*.

The small creek meandering through the swampy marsh area was the camp water supply. However, the creek and marsh also served as the prison latrine. They were filled with fecal matter from thousands of sick and dying men. With no other source of water, POWs had no choice other than to drink from the swamp and polluted creek.

POWs at Andersonville were akin to a primitive society. Survival often depended upon a prisoner's social network skills. A prisoner with caring friends was more likely to survive, for friends offered shelter, food, blankets, and emotional support. Loners withered and died. During the 14 months that Andersonville was in operation about 44,000 prisoners passed through its gate. The highest POW count at any one time was 31,693 men on 31 August 1864.

Over 12,000 prisoners died in the prison. Overcrowding, filth, disease, exposure to the elements, and lack of food and medicine all contributed to an overall mortality rate of 28 percent. The camp commander, Capt. Henry Wirz, did what he could with the meager assets at hand. In July 1864 he prepared a multi-page petition that was signed by most POWs then in the camp. The petition asked the Union to reinstate prisoner parole and exchange to relieve gross overcrowding. Wirz paroled five POWs who agreed to carry the petition north to the U.S. Army, which then was besieging Atlanta. The POWs delivered the petition, but the request was denied.

One POW at Andersonville had been attached to the 154th New York Volunteer Infantry. Initially he had been imprisoned at Belle Isle prison camp in Virginia. In early 1865 the Confederacy moved him to Andersonville. He survived in southern prison camps for a total of 661 days. For the record books he is credited with being the longest-held Union POW.

When a POW at Andersonville died his fellow inmates stripped him and saved his clothing and anything else of conceivable value. They then tied a tag to the corpse's big toe, documenting the dead man's name and military unit. They carried the corpse to the main gate and left it there. Once each day, contract laborers would take all corpses outside the stockade for burial.

> Death is so commonplace it doesn't shock you anymore.
> *[LtCol. H. Lee Bell, USMC, 1369, 1992]*

Many months before the war ended the northern press began

printing sensational and usually embellished articles about horrors at Andersonville. This led many people, including some officials in the U.S. Government, to believe that Union POWs were deliberately mistreated and starved. The public clamor for vengeance could not be ignored. Northern newspapers demanded retribution. Someone had to be held accountable.

POWs at Andersonville were liberated in May 1865, the month after the South capitulated. Federal troops arrested Capt. Wirz. He was charged with (1) eleven counts of murder and (2) conspiracy to destroy the lives of federal prisoners. In the first war crimes trial in United States history, he faced a military tribunal in Washington. Several ex-prisoners testified that they saw Wirz shoot prisoners. Others testified that he ordered guards to beat prisoners mercilessly. Wirz quickly was found guilty and sentenced to death. On 10 November 1865 he was hanged at Old Capitol Prison. One noted historian, George Rugg, later would write:

> *[Wirz]* was guilty of neither of the charges. Testimony gathered by the prosecution was in some instances exaggerated, *[and in other instances]* wholly fabricated.

Later it was found that Wirz was not at the prison when many of his alleged shootings took place. Further, some POW witnesses had been in other prisons at the times they claimed to have been beaten at Andersonville. Several documents produced by the prosecution were found to have been forged and falsified.

However, regardless of the cause, claims of inhuman conditions at Andersonville were well founded. The 28 percent mortality rate speaks for itself. Overcrowding was the prime culprit. At its peak the POW count was nine times the designed capacity of the prison. Extremely limited rations of poor quality, lack of clothing, lack of shelter, contaminated water, absence of medical care, and rampant contagious diseases all combined to take a lethal toll. One survivor wrote that he could walk around looking for acquaintances and "see a dozen almost any morning, laying around dead." George Rugg explained:

> There were simply too many prisoners and not enough food, clothing, medicine, or tents to go around. . . . Shortage of food throughout the Confederate states, and

> the refusal of Union authorities to reinstate the prisoner exchange, are also cited as contributing factors.

Soldiers endured great perils during battle. Nonetheless, those who were captured found prison camps to be far more dangerous than combat. POWs who survived in captivity often had both physical ailments and psychological scars that lasted for a lifetime.

Many thousands at Andersonville had failed to survive. In later years the camp burial ground would become Andersonville National Cemetery. It contains graves of 13,714 Union soldiers. Of those, 921 are marked "unknown."

Elmira Prison Camp: In western New York at the town of Elmira, Camp Rathbun had been a Union Army training base and muster site. In the summer of 1864 the Union converted part of the camp into a military prison to house Confederate POWs. The Camp Rathbun name fell into disuse, and the facility would become known as Elmira Prison Camp. It was the most infamous prison camp in the Union. Of the approximately 11,200 prisoners confined there, 2,963 would die in captivity. Survivors would call the camp "Hellmira."

To make the former training base suitable for housing prisoners, workmen built a wooden stockade wall around the facility. The wall was completed in June 1864 and encircled 35 wooden barracks that were intended to house up to 4,000 inmates. The new prison opened on 6 July when the first train load of 399 POWs arrived. More came each day, and by the end of the month over 4,400 prisoners had been confined in the camp. During the next month the POW tally soared to 9,600. The earliest arrivals, about half of the total, had been assigned sleeping space in the barracks. The remainder were housed outside in small canvas A-tents, three men to a tent. Prisoners wore ragged summer clothing, and they had no blankets.

Food initially was adequate. However, on 18 August the camp commander, Col. William Hoffman, put POWs on bread-and-water rations. This change, he explained, was in retaliation for reported mistreatment of Union POWs in southern prisons. The new diet caused rapid onsets of scurvy, and 1,870 cases had been reported by 11 September. By the end of the year 1,264 POWs had died. That winter they succumbed at the rate of 25 per day from scurvy, diarrhea, exposure, pneumonia, smallpox, and starvation. Not until December had the bread-and-water ration been supplemented with meat.

> This morning early, there were 18 dead bodies lying naked on the bare earth, *[and]* more were added to the list by half past 8 o'clock.
> *[Anthony Kelly, diary entry on 21 September 1864]*

The winter of 1864-1865 had proved to be unusually bitter. Each barracks held 200 prisoners and had two small wood-burning stoves. Thus, each stove had been the only source of heat for 100 men. To compound the dilemma a very limited amount of wood had been furnished for use in the stoves. There was not enough wood to keep a fire burning all day or all night. Prisoners suffered horribly from the cold. Families with relatives or friends imprisoned at Elmira shipped thick clothing to them. Few prisoners ever had the chance to wear it. One historian later would explain:

> Hoffman would allow only items *[of clothing]* that were gray to be distributed. Clothes of other colors were burned, while the sons and husbands for whom they were intended literally froze to death.

The plight of prisoners in the barracks paled in comparison to conditions suffered by those outside in tiny A-tents. Even worse, the supply of tents had run out on 7 August. Thereafter new arrivals had to sleep on the ground with no shelter whatsoever from rain and winter snow. Such prisoners often vied for space under A-tents. It was common to find five or more skeletal prisoners jammed under one little ragged tent.

Rations also had gone downhill. Even before the start of a bread-and-water diet, the camp kitchen had been unable to cope with the thousands of prisoners. After preparing meals for 4,200 POWs, the rest of the prisoners had gotten their food raw. For sick prisoners the camp had a hospital tent, but rarely was a doctor there. By November the death toll recorded at the hospital tent had reached 755 men. By the end of December, 200 dying men lay in their tents or barracks because there was no room for them in the hospital tent.

Elmira had no uncontaminated water supply. On the southern side of the camp was a pool called Foster's Pond, a backwash from the Chemung River that flows nearby. Calculated on the basis of 10,000 POWs, the U.S. Sanitary Commission reported that about 2,600 gallons of prisoner urine was released into Foster's Pond each day. Worse yet, latrines used by thousands of POWs had been dug adjacent to the pond. Camp garbage was deposited in the same area. Nonetheless, it was the only water source for prisoners. The pond and surrounding vicinity became a breeding ground for disease. One prisoner described the cesspool as "green with putrescence and filling the air with messengers of death."

> *[Prisoners]* fished in the pond until the water turned green with filth. Rats ran rampant in the camp, and killing them with rocks became a sport. Men skinned the rats *[to make]* gloves; then starving men boiled and ate the rest.
> *[Sgt. Berry Benson, CSA, diary entry, 1864]*

Sgt. Thaddeus Davis of the 40th North Carolina Regiment had been captured at Fort Fisher, North Carolina, on 15 January 1865. He was imprisoned at Elmira, and many years later he would write:

> We arrived at *[Elmira]* about eight o'clock in the evening in four feet of snow. Many prisoners had neither blankets nor coats. . . . Those who preceded us were . . . dirty, pale, emaciated, and scantily clothed. Our rations consisted of a *[slice]* of stale bread an inch thick. . . . When prisoners died their bodies were stacked up in a dead-house as high as they could stack them. . . . *[Guards]* were the meanest men I ever saw, demons in human flesh.

For the prisoners a new indignity arose. Civilian entrepreneurs built a high observation platform outside the stockade wall. It was equipped with chairs and binoculars. One historian explained:

> Visitors were charged 10 cents apiece to use the platform and look at the prisoners. Refreshments were sold to viewers while Confederate prisoners starved.

In the absence of meat and vegetable rations, prisoners resorted to trapping rats that infested the camp latrine area. From whatever source, fresh meat presented POWs with a welcome treat. Rat meat became a bartering staple in the camp. Hoffman did not approve of this new meat source, so he obtained a small black dog to catch and kill the rats. Two POWs from North Carolina caught, killed, cooked, and ate the dog.

Talk of escape was the main topic on the lips of POWs. Some did more than just talk. Washington Traweek and Berry Benson led a group of prisoners who began digging tunnels. One was discovered, but the other was not. It began in an A-tent and burrowed under the stockade wall. Traweek, Benson, and eight more men escaped during the dark of night and eventually made their way to freedom.

The number of dead and dying men rocketed upward by January 1865. The winter weather was freezing cold. Some new barracks had been built, but hundreds of POWs still huddled under tents outside. They battled filth, disease, the cold, and starvation. At morning roll call some stood ankle-deep in snow with rags tied around their frostbitten and swollen feet. The U.S. Sanitary Commission begged for permission to send medical personnel to aid critically ill prisoners. Hoffman denied the request. Of the 11,200 Confederate POWs confined at Elmira, 26 percent would die of hypothermia, various illnesses, and starvation.

The war ended in early April 1865, and the camp began the parole process within a week. As a condition of parole each POW was required to sign the following document under oath. Those unable to write could "make their mark" and have it witnessed:

> – Amnesty Oath –
>
> I,____*(printed name)*____, of ____*(hometown)*____, do solemnly swear, in the presence of Almighty God, that I will henceforth faithfully support, protect, and defend the Constitution of the United States and the Union of the States thereunder, and that I will in like manner abide by and faithfully support all laws and proclamations which have been made during the existing rebellion with reference to the emancipation of slaves, so help me God.
>
> ______*(signature)*______

Upon release from Elmira surviving former prisoners were given two meal rations, ten dollars, a Union railroad voucher, and were put on a train bound for City Point, Virginia. One former prisoner would write in later years:

> Out *[of the prison camp]* they came, a ghastly tide, with skeleton bones and lusterless eyes, and brains bereft of but one thought, and hearts purged of but one feeling – the thought of freedom, *[and]* the love of home.

City Point was a major U.S. Army supply depot. From there the newly freed men would have to straggle home by whatever means might be available. The southern railway network was more or less inoperative. Consequently, after detraining at City Point most men would have to start walking. Pvt. Henry Evans was one of the men who had to walk all the way home. Upon arrival he was so emaciated that his mother, sister, and brother could not recognize him.

One stubborn prisoner from South Carolina, Pvt. William Sturkey of Company B, Hampton Legion Regiment, had refused to sign the amnesty oath. He told his captors that, as a matter of honor, he could not renounce the vow he had taken when he joined the Confederate Army in 1861. Therefore, he was not released. Three months later the hard-headed POW was still in Elmira prison, and he still refused to sign the oath. He was given the option of taking the oath verbally, and he refused. On 3 July 1865 exasperated federal authorities relented and told him simply: "Go home!"

The defiant rebel soldier made his way to the rural village of Dorns Mines (later, *McCormick*) in South Carolina, where he lived for the rest of his life. When he died 56 years later in 1921 his lengthy obituary in *Confederate Veteran* magazine would conclude as follows:

> . . . He lived, died, and was buried a Confederate Soldier, dressed in the Confederate Gray, one who stayed in a Northern prison three months after the surrender because he refused to take a prescribed oath renouncing the Confederate cause.

On 5 July 1865, two days after Pvt. Sturkey had been released, Elmira Prison Camp stood vacant. No possible future usage was foreseen, so the U.S. Army demolished the camp.

Pvt. William Kesler has the dubious distinction of being the last Confederate soldier to leave the state of New York. While attached to the 8th North Carolina Infantry Regiment he had been captured at Gaines Mill, Virginia, and imprisoned at Elmira. After the war ended

Kesler took the required oath and was paroled, but he was considered too sick to survive the train ride. Civil authorities took him to the nearby Union Army hospital at Camp Chemung. He stayed there until 11 September 1865, when he was deemed fit to board the train. Kesler safely reached his home in Salisbury, North Carolina, but he never recovered from privations suffered in the prison. He died less than three years later on 5 June 1868. He is buried in Union Lutheran Church Cemetery in Salisbury.

In 1876, eleven years after the war ended, *Southern Historical Society Papers, Volume 1*, would include a report from a U.S. Army medical officer who had served at Elmira. In part the report reads:

> I found, on commencement of my duties at Elmira, about 11,000 rebel prisoners *[with]* diseases owing to an improper diet, a want of clothing, necessary shelter, and bad surroundings; diseases were consequently of the following nature: Scurvy, diarrhea, pneumonia, and the various branches of typhoid. . . . The winter of 1864-65 was an unusually frigid one, *[and]* this alone was ample cause for an unusually high mortality *[rate]* amongst them. . . . A turbid stream of water, horrible to relate, was the only source of *[water]* the prisoners could use. . . . Hence, it is readily to be seen that the range of mortality was no less at Elmira than at Andersonville.
>
> At Andersonville there was actually nothing to feed or clothe the prisoners with. *[Confederate]* soldiers fared but little better than their prisoners. *[At Elmira]* a slice of bread . . . was given *[to prisoners]* for breakfast, a poor hatch-up concocted cup of soup, so called, and the slice of miserable bread. . . . Hundreds of sick *[men]*, who could in nowise obtain medical aid, died – unknelled, unconfined, and unknown. The facts *[above]* can be attested by the staff of medical officers who labored at the Elmira prison for rebel soldiers.

In 1907 many years after the war had ended the old prisoner burial ground adjacent to Elmira Prison Camp would be designated as Woodlawn National Cemetery. The U.S. Government replaced all old grave markers with marble headstones.

John A. King, a POW, had arrived at Elmira on 1 August 1864.

He had been paroled in early June 1865 and had made his way to his home near Clarksburg, West Virginia. Fifty-two years later he began writing memoirs of his life in Elmira Prison Camp:

> No battle scarred veterans ever marched to victory prouder than that ragged, poorly fed, miserable 300 which passed out of the main gate, never to return. . . . Dear old comrades in misery, how often do I remember you and your friendship! . . . The people at home never knew how we suffered in prison. . . . I am proud to say that I once belonged to the Army of Northern Virginia and marched and fought under the illustrious Robert E. Lee. We started out for what we thought was right and stayed with it faithfully to the bitter end.
> *[John A. King, CSA, in his memoirs, 1917]*

> There is many a boy here today who looks on war as all glory. But, boys – war is Hell!
> *[Gen. William T. Sherman, USA, speaking to military veterans and young men, 12 August 1880]*

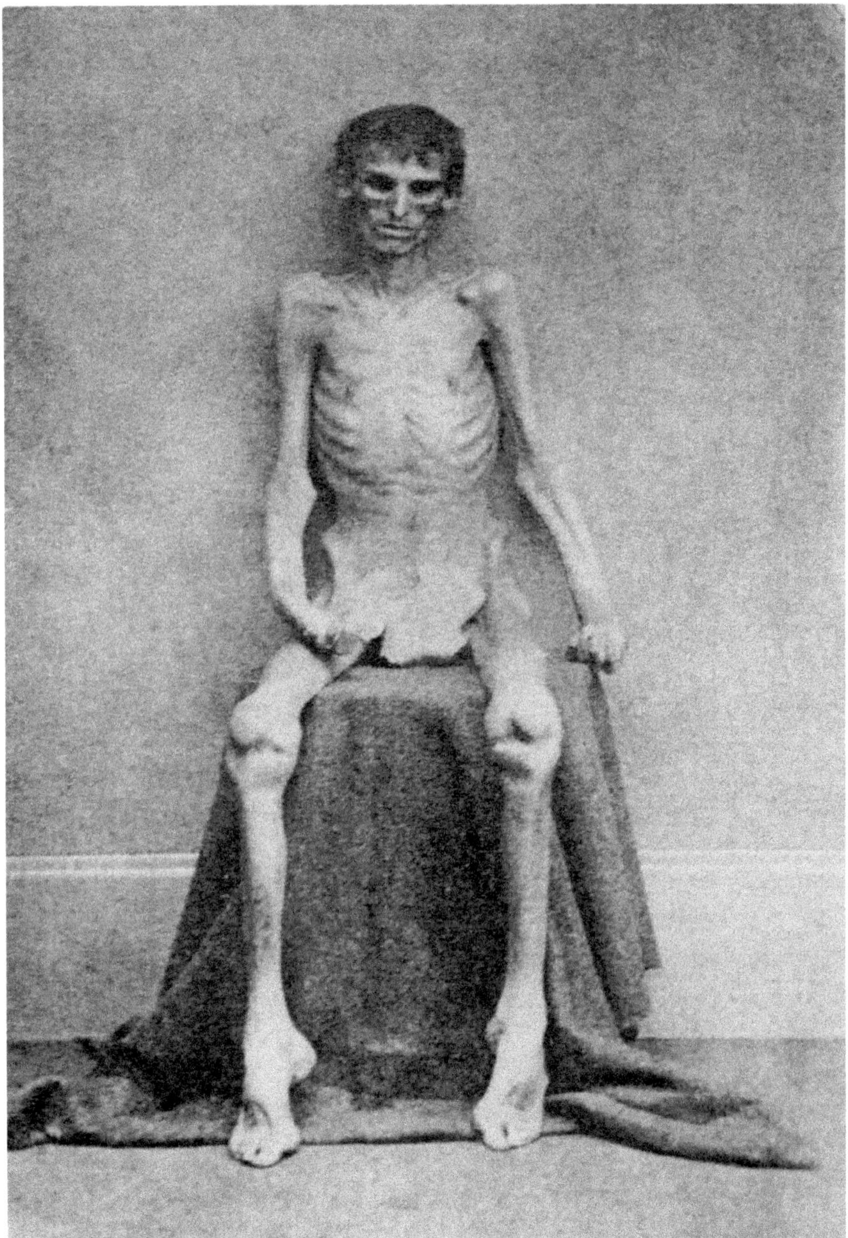

A former prisoner of war, photographed upon his release in 1865.

Photograph of A-tents for prisoners of war at Elmira Prison Camp.

Confederate Memorial at Woodlawn National Cemetery. The national cemetery is the former burial ground for Confederate prisoners who died in Elmira Prison Camp.

Image of A-tents for prisoners of war at Andersonville Prison Camp.

Image of one of the burial trenches for prisoners of war who died while confined in Andersonville Prison Camp.

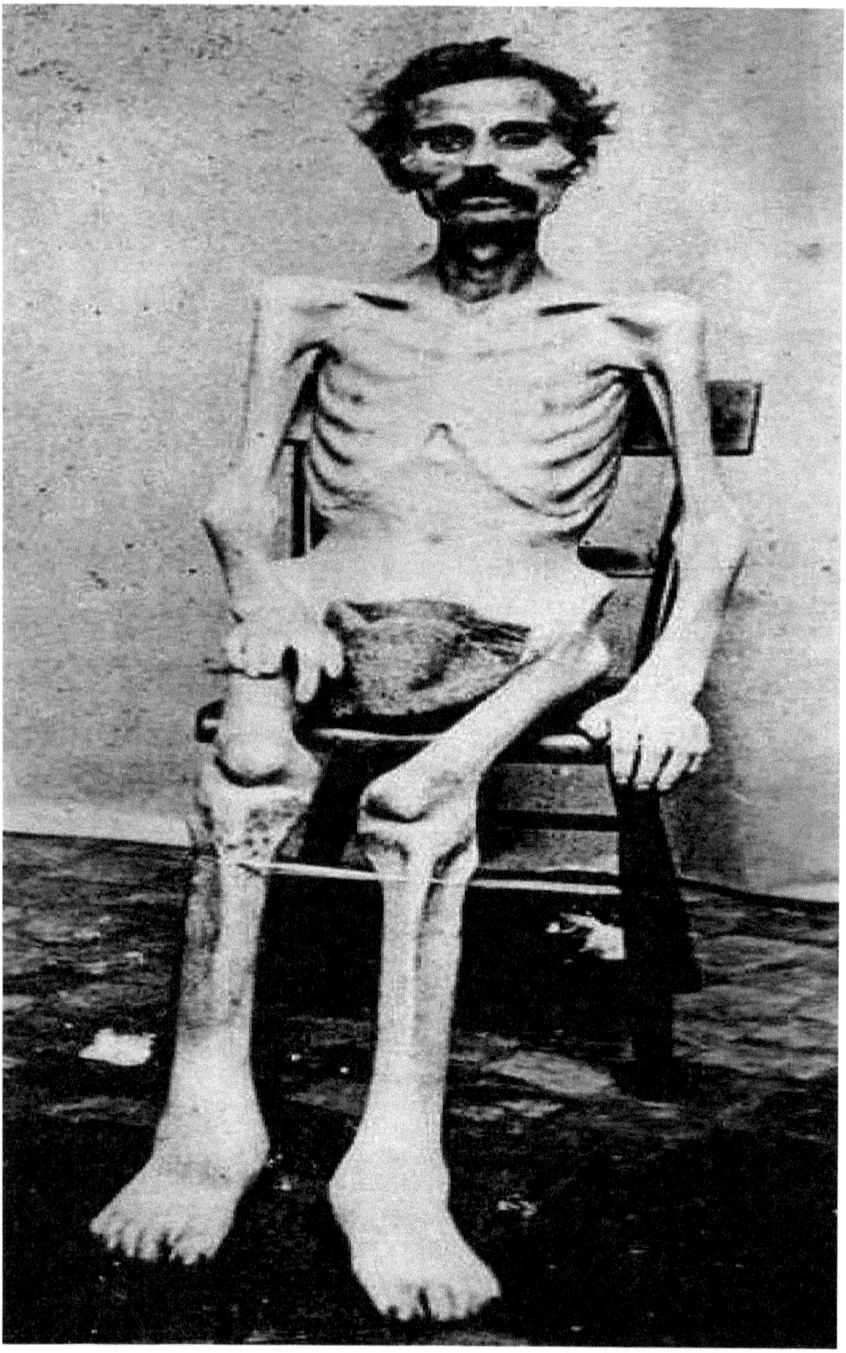

A former prisoner of war, photographed upon his release in 1865.

Letters, Diaries, and Memoirs

> The old South has passed away; her music is all dead; her harp *[is]* hung where no mortal hand can sweep its chords again. The very wind of Heaven can bring from it naught save a few wailing notes . . . .
> *[Judith W. McGuire, preface to Diary of a Southern Refugee during the War, 1867]*

Letters, Diaries, and Memoirs: Soldiers wrote letters to family and friends during the war. Some kept diaries, and others later would chronicle their wartime experiences in memoirs. From such letters and documents one can get a sense of how soldiers lived and the horrors they faced in battle. Also, letters and diaries from the home front offer insight into southern society during the long conflict.

For Confederate soldiers, letters from home were a pleasure and a treasure. Further, many packages from their families contained clothing: hats, coats, shirts, trousers, and invaluable shoes. Often pies and cakes were included, as well as hard candy and other foodstuffs that would resist spoilage in transit.

The Union Army maintained post offices near its forts and camps, and mail service followed the troops. Initially Union soldiers had to purchase their own envelopes (called, *covers*), paper, and stamps. Later the U.S. Sanitary Commission and U.S. Christian Commission distributed paper and envelopes to Union troops free of charge. By 1864 the U.S. Mail began to deliver mail from soldiers without the need for postage stamps. To avail himself of this benefit a soldier would write "Soldier's Mail" on the face of the envelope.

There were no such benefits for Confederate troops, and they experienced a shortage of envelopes, paper, and stamps. Each man had to purchase these items as well as pencils, or pens and ink. Often mercenary vendors (called, *sutlers*) visited Confederate camps and offered these items at a reasonable price.

In both armies some soldiers, in particular those from rural areas, could neither read nor write. Literate comrades would write letters for them and read to them any letters from home. On the other hand, most soldiers from both the North and South had gotten at least *some* formal schooling. The average young combatant could read and write at a fourth grade level. At the extremes, some soldiers were highly educated. Conversely, lack of education was evident in letters that featured a host of misspelled words and grammatical flaws.

Letters, diary entries, and memoir excerpts below are listed in random order. Most are from Confederate soldiers and their families, but a few are from Union troops. In the interest of clarity, spelling and grammatical errors have been corrected without notice to the reader. Explanatory interpolations have been added in the interest of readability and clarity.

****

From Fredericksburg, Virginia, a Confederate sergeant in the 32nd Virginia Infantry Regiment wrote to his wife in November 1862. From her recent letter he had learned that his last four letters to her had not been received. In his letter he related his involvement in the brutal Battle of Sharpsburg (aka, *Battle of Antietam*) on 17 September 1862, the single bloodiest day in American history:

> My dear and devoted Mattie,
>
> . . . that memorable battle of Sharpsburg! Oh, Mattie, it almost makes me shudder to think of how the 17th of Sept. '62 will live in the memory of the 32nd Va. Regiment & its friends. Oh, it was an awful day. Imagine my feelings, if you can, when I saw my comrades and friends falling all around me from the death-dealing shot and shell of the enemy and knew not how soon it would be my fate. Thanks to a kind Providence I came out unhurt, but narrowly escaped. A ball passed through my blanket between my body and right arm. I shot my gun until I could hardly get a cartridge down her. Finally they gave way & ran, pursued by our brigade.
>
> Dearest, I know you do not want to see me any worse than I, for you are the joy and pride of my life. It is you and little *[illegible name of a child]* that make life dear to me now. Dear Mattie, I must tell you I am in a hard place and know not what to do. I think of my sufferings both in mind and body, which are indescribable, and how much my services are needed at home . . . .
>
> The war is likely to last for years yet, and I cannot

reasonably expect to survive it. . . . Give my love to all, and may the good Lord soon bring us to meet again in the flesh is my humble prayer and heart's desire.

Your affectionate husband,

==========

From a camp near Chattahoochee, Georgia, a Confederate soldier wrote to his sister in July 1864. Although elements of the Union Army were camped nearby on the other side of a river, there had been no recent fighting between the two forces:

Dear sister,

. . . Pickets don't fire at each other now. We go down to the edge of the river on our side, and the Yankees come down on their side, and *[we]* talk to each other. The men on picket duty *[across the river]* are from Ohio and seem very tired of the war.

The Yankees are very much in want of tobacco, and our Government gives it to us, and we . . . trade tobacco with them for knives and canteens. There is a rock near the middle of the river to which they would swim and trade. After a while they got so well acquainted that some of our men would swim across and land among the Yankees.

The other day as I was sitting on the *[river]* bank one of the Yankees from the other side called to me to *[ask]* if I had any tobacco. I told him I had. He said that he had a good knife to trade for it, *[and he added]*, "I want a chew of tobacco very bad."

Write to me soon,

==========

From his camp near Bunker Hill, Virginia, a Confederate soldier from the 4th North Carolina Volunteer Infantry Regiment wrote to his family in early September 1862 before the coming firestorm at Sharpsburg. He and his regiment were in the Confederate Army that had invaded Maryland:

Dear Folks,

On the 8th *[of September 1862]* we struck up the

> refrain of "Maryland, My Maryland" and camped in an apple orchard. We went hungry; not a morsel of bread or meat had gone into our stomachs – and our menu consisted of *[only]* apples and corn. We toasted, we burned, we stewed, we boiled, we roasted these two together, and singly, until there was not a man . . . who had not *[had]* a bad attack of diarrhea. Our underclothes were foul and hanging in strips, our socks worn out, and half of the men were barefooted . . . .
>
> The command continued the march through the rich fields of Maryland. The country people lined the roads, gazing in open-eyed wonder upon *[our]* long lines of infantry . . . as far as the eye could reach was the glitter of the swaying points of bayonets. It was the *[illegible]* ragged rebels they had ever seen . . . and they gave *[food]* liberally, and every haversack was full that day at least. No houses were entered – no damage was done. Farmers in the vicinity must have drawn a long breath as they saw how safe their property was in the very midst of *[our]* army.
>
> Warmest wishes,

Eight days later on 16 September, the day before the pivotal battle at Sharpsburg, the same soldier sent another letter to his family. This time he asked his father to send him a pair of shoes:

> Dear Father, Mother, and Sisters,
>
> It has been some time since I wrote to you all. I have heard from you two or three times. I have been in Maryland since I wrote to you and have been in two very hard battles . . . and came out unhurt.
>
> . . . Pa, I want you to have me a pair of shoes made. Those shoes you had made for me ripped all to pieces. Our regiment used everything we had. I have no blanket nor any clothes *[except what I am wearing]*. I have got the suit that you sent me. *[It]* came in good time. I like *[it]* very well. If I had a good pair of shoes I would be the best clothed man in the regiment. . . . I have nothing more

[I can] write, for my paper is scarce.

Write soon to your only son,

==========

A young Confederate lieutenant from Company C, 8th Tennessee Cavalry, wrote to a girl that he knew well. His letter was postmarked in Gallatin, Tennessee, on 25 July 1864:

> Dear Miss,
>
> It was a pleasure to me to have the honor to receive a letter from as charming a young girl as *[you]*. . . . The letter that you wrote to me has proved very satisfactory to me. If you will stand up to what you told me in your letters I will be satisfied. . . . You are the girl that I am depending upon. . . . Though we are far apart at present, my heart is with you every moment. . . . The thought of your sweet smile is all the company I have. I trust that you are sincere in what you have written me.
>
> Your love and friend,

==========

Near his field hospital at Antietam (aka, *Sharpsburg*) a Union surgeon with the 5th New Hampshire Volunteer Infantry Regiment wrote to his wife. His letter was dated 22 September 1862, five days after the Battle of Antietam (aka, *Battle of Sharpsburg*):

> My Dear Wife,
>
> The days after a battle are a thousand times worse than the day of the battle – and the physical pain is not the greatest pain suffered. How awful it is – you have not, until you see it, any idea of affairs after a battle. The dead appear sickening, but they suffer no pain. But the poor wounded mutilated soldiers that yet have life and sensation make a most horrid picture. I pray God may stop such infernal work – though perhaps He has sent it upon us for our sins. Great indeed must have been our sins if such is our punishment.
>
> I have made the acquaintance of two *[wounded]* rebel officers – prisoners in our hands. One is a physician – both

are masons; both *[are]* very intelligent gentlemanly men.

Kisses to you, Clint, and Kate,

==========

A Union soldier with the 14th Connecticut Volunteer Infantry Regiment made the following entry in his diary on 17 September 1862. Earlier that day at Antietam (aka, *Sharpsburg*) he had fought near the infamous Sunken Road:

> This has indeed been a fearful day, and it is by God's kindness alone that I am here to write this. We woke up early in the morning. I went out and read the Bible and a prayer. In a few minutes the enemy began to throw shells at us *[and]* the shells burst all around us and *[among]* us. . . . One shell burst so near as to scatter dirt in my face as I lay upon the ground. *[In the evening]* as we rested for the night, Company A *[could]* muster only 6 men besides the captain. . . . Thus ended the first day of the battle, and a fearful one it was.

==========

Quoted in *Battles and Leaders, Volume II*, a Union private with Company G, 9th New York Volunteer Infantry Regiment, described the customary reaction to heavy enemy fire:

> The truth is, when bullets are whacking against tree trunks, and solid shot are cracking skulls like eggshells, the consuming passion in the breast of the average man is to get out of the way.

==========

A Union private with Company K, 125th Pennsylvania Volunteer Infantry Regiment, described the aftermath of the titanic struggle at Antietam (aka, *Sharpsburg*) in the scrapbook of a friend:

> Under the dark shade of a towering oak near the Dunker Church lay the lifeless form of a drummer boy, apparently not more than 17 years of age, flaxen hair and eyes of blue and form of delicate mold. As I approached him I stooped down, and as I did so I perceived a bloody

mark on his forehead. . . . It showed where the leaden messenger of death had produced the wound that caused his death. His lips were compressed, his eyes half open, a bright smile played upon his countenance. By his side lay his tenor drum, never to be tapped again.

==========

A Confederate colonel was quoted in *Lee's Lieutenants*:

Such a storm of balls I never conceived it possible for men to live through. Shot and shell shrieking and crashing, canister and bullets whistling and hissing most fiend-like through the air until you could almost see them. In that mile's ride I never expected to come back alive

==========

A war correspondent for the *Boston Journal* newspaper was quoted in *Battles and Leaders, Volume II*:

It was no longer the boom of the *[artillery]* batteries, but a rattle of musketry – at first like pattering drops upon a roof; then a roll, crash, roar, and rush, like a mighty ocean billow upon the shore, chafing the pebbles, wave on wave, with deep and heavy explosions . . . like the crashing of the thunderbolts.

==========

A woman known only as Ettie received a letter dated 5 January 1864 from Nashville, Tennessee. From the letter content, Ettie is the friend or relative of an unidentified soldier, and their relationship may have been – or perhaps *may not* have been – platonic:

Friend Ettie,

If I get out of the service alive I am going to settle down and get married. What a novel idea that is. Perhaps you will not believe it, but I am not joking. I am not quite an old bach*[elor]* yet, but I fear I will be before long.

If you know some good-looking amiable young lady that wish*[es]* to change her situation in life, just mention the fact to her, and tell her there is a soldier in the army that wishes to marry in less than two years after his time expires in the army . . . .

Yours very truly,

==========

A Union lieutenant with Company H, 9th New York Volunteer Infantry Regiment, lying prone under fire, described an order to stand erect and form his men in line for battle:

> I was lying on my back, supported on my elbows, watching the shells explode overhead and speculating as to how long I could hold my finger up before it would be shot off, for the very air seemed full of bullets. When the order to get up was given I turned over quickly to look at Col. Kimball, who had given the order, thinking he had become suddenly insane.

==========

A Virginia Military Institute (VMI) cadet had been killed in battle in May 1864. The C.S. Army had notified the young cadet's family. Thereafter, VMI sent the dead cadet's father the following letter:

> Sir,
>
> You have doubtless received before this the mournful intelligence that your son has been added to the long list of the gallant dead who have fallen in defending their country against the invasion of a ruthless foe. . . . You also have received . . . an official letter from the Adjutant informing you of the sad event.
>
> I can add nothing more except . . . that it may be concluded that his death was instantaneous. This *[information]* I received from Col. Gilham, who examined the body before its interment. The quartermaster will endeavor to preserve any mementoes or any property of the cadets who have fallen . . . .
>
> I *[can]* offer no words of condolence. *[Yet]*, I know how to sympathize with you, for my noblest son fell slain in battle not two months after he left the Institute – and I know by experience that the only comfort for so great a sorrow must come from a source higher than any on earth.

==========

On 23 June 1864 a Confederate soldier penned this letter to his family in Mississippi:

> Dear wife, mother, sisters, and children,
>
> I am well, all but my arm. The sore on my arm is nearly cured up, but the rising under my arm is swollen very much. It is as large as your two fists.
>
> . . . I put my trust in God, and if He is willing I think I will be permitted to see you and *[our]* family again. God is a just and all wise being. . . . My prayers ascend to God on your behalf. Hereafter I hope to meet God at the judgement and then be welcome to a seat on his right hand. And it is my prayer that you may be prepared to go with me, there to dwell in peace & happiness, where there is no war and no pestilence.
>
> I have prayed to God earnestly to let me live to see this unholy war close, and live to see my little children raised until they are all grown and able to take care of themselves. . . . I think I will see you before long, for I do want to see you so bad. I long to see my little home again.

==========

On 4 July 1861, national Independence Day in the United States, a Confederate soldier in the 2nd Louisiana Infantry Regiment made the following entry in his diary:

> Camp *[is]* in a fever of excitement on account of rumors . . . a battle is expected in a day or so. This is the first 4th *[of July]* dawning on our glorious Confederacy, but what a change has come upon the country, "the home of the free and the land of the brave *[sic]*." We are a disrupted union, & I feel sensibly that, after all, no written constitution & laws framed by the wisest of our ancestors, no common country made sacred by the blood of our forefathers, no brotherly share in past victories is *[worth the present separation]* that wreaks its passion in slaughter . . . . The best of human governments is a failure & was but an experiment.

==========

From Union lines at Cold Harbor, Virginia, a Union sergeant in the 89th New York Infantry Regiment wrote a letter to his wife on 6 June 1864. That week at Cold Harbor, Union regiments suffered perhaps the most lopsided defeat of the war. They made repeated frontal assaults against entrenched Confederate infantry, supported by artillery, atop a ridge. Each Union charge had been repulsed with extremely heavy losses:

> My Dear *[wife]*,
>
> I am still among the living, but still I have to acknowledge that life here at present is very uncertain.
>
> There are two whole lines of rifle pits between ours and our pickets, and behind ours *[are]* at least six or eight more, and the troops are massed in all of them in two ranks, so if the rebels attempt to break through here they will have a hard road to travel.
>
> The rebel works are very strong in our front, and it will be impossible to take them by storm. The night before we came here there was a charge made on them, but our men were repulsed, and the boys that were on picket say the ground is almost covered with our dead, and they can't be *[recovered]*. Our pickets & the rebels are not over a dozen rods *[a "rod" equals about 16 feet]* apart and an almost constant fire is kept up.
>
> Yours ever,

Two months later on 11 August 1864 the same Union sergeant was in charge of security guards and pickets at Petersburg, Virginia, where Union troops again faced daunting Confederate defenses. The sergeant wrote a letter to his wife:

> Dear wife,
>
> On several occasions during this campaign I have dreamed of, and told the *[soldiers about]*, events that really did transpire. A spell ago one of the sharpshooters of our Corps dreamed that he was to be killed the next day. In the morning he told the sergeant, and *[he]* wanted to be excused from *[duty]* that day, for he *[had dreamed that]*

he would surely be killed. But the sergeant said there was nothing in dreams, and that he *[would not be excused from duty]*. And sure enough, before nine o'clock he stood up to look toward the enemy trenches, and a rebel shot him through the head, killing him instantly.

Accept the love of your affectionate husband,

==========

A Confederate soldier in the 1st South Carolina Infantry Regiment wrote a letter to his mother on 28 June 1863. At that time, before the Battle of Gettysburg a week later, the South had invaded the North and was camped near Chambersburg, Pennsylvania:

My Dear Mother,

We crossed the Pennsylvania and Maryland river yesterday morning and traveled about 12 miles into this rich country. Today we have rested and have endeavored to observe the 4th Commandment [meaning, *"Remember the Sabbeth day and keep it holy"]*. The whole army is now together and is striking towards the very heart of Pennsylvania. . . . This is a grand expedition, and I trust it will be fruitful of good results to us.

We have been pressing splendid horses into service. . . I have seen the largest & finest horses bought here for 100 & 125 dollars. Cattle and sheep & etc. will also be pressed. The people in most instances have not refused to take our money, and we have bought chickens, butter, & etc. in abundance at fabulously small rates. I went into a store in a small town yesterday & bought sugar at 20 cents.

The people are afraid not to take our *[Confederate]* money – they are completely subdued & are as polite & civil as possible, longing for peace. I trust the hour of delivering it *[is]* at hand and, God assisting us, we may conquer a peace this summer.

Your loving son,

==========

On 14 July 1861, a week before the First Battle of Bull Run (aka, *First Battle of Manassas*) on 21 July, a major in the Union Army wrote a letter to his wife. Perhaps he had a premonition of events to come, for his written words foretold his death. He was, in fact, killed in battle at Bull Run seven days after he penned this letter:

> My very dear Sarah,
>
> Lest I should not be able to write to you again, I feel impelled to write lines that may fall under your eye when I shall be no more.
>
> . . . Sarah, my love for you is deathless. . . . The memories of the blissful moments I have spent with you come creeping over me, and I feel gratified to God and to you that I have enjoyed them so long. And hard it is for me to give them up and burn to ashes the hopes of future years when, God willing, we might still have lived and loved together and *[have]* seen our sons grow up to honorable manhood around us.
>
> I have, I know, but few and small claims upon Divine Providence, but something whispers to me – perhaps it is the wafted prayer of my little Edgar – that I shall return to my loved ones unharmed. If I do not, my dear Sarah, never forget how much I love you, and when my last breath escapes me on the battlefield, it will whisper your name.
>
> . . . Forgive all my faults and the pains I have caused you. How thoughtless and foolish I have oftentimes been! How gladly would I wash out with my tears every little spot upon your happiness, and struggle with all the misfortune of this world to shield you and my children from harm. But I cannot. I must watch you from the spirit land and hover near you, while you buffet the storms *[of life]* and wait with sad patience 'till we meet to part no more.
>
> But, O Sarah! If the dead can come back to this earth and flit unseen around those they loved, I shall always be near you; in the garish day and in the darkest night – amidst your happiest scenes and gloomiest hours – always, always. And if there be a soft breeze upon your cheek, it shall be my breath; or *[if]* the cool air fans your throbbing temple, it shall be my spirit passing by.
>
> As for my little boys . . . little Willie is too young to remember me long, and my blue-eyed Edgar will keep my frolics with him among the dimmest memories of his child-

hood. Sarah, I have unlimited confidence in your maternal care and your development of their characters. O Sarah, I wait for you! Come to me, and lead thither my children.

Yours,

==========

A Confederate surgeon wrote a lengthy letter to his wife on Sunday night, 1 November 1863. The letter was written at Brandy Station, Virginia:

My dear Mollie,

I received a letter today from a handsome lady. . . . I first met her when I was about 17 years old and she was 8. I afterwards saw her frequently and occasionally was in her company, and notwithstanding the disparity of our ages, I became so favorably impressed with her fair face and gentle manner that I frequently said to myself that I wished she was older, or I younger.

. . . *[Years later]* the happy day of our marriage arrived. . . . How happy the thought that years increase the affection & esteem we have for each other, to love & be loved. May it ever be so, and may I ever be a husband worthy of your warmest affections . . . .

Yours affectionately,

==========

A Confederate soldier from the 1st Texas Volunteer Infantry Regiment wrote to his father on 28 November 1862. The letter likely was written somewhere near Maryland:

Pa,

I wrote you soon after the fight & gave you all the information I could about Robert. I have been inquiring and hunting for him ever since he was lost. I can hear nothing from him. I feel that he was slain, although I cannot give him up yet. There is some chance for him to be alive yet. He may have been badly wounded and still in the hands of the enemy. There have been some of my boys sent back . . . that I thought were killed. They saw nothing of Robert,

> but say he may be there somewhere, as our boys are scattered all over Maryland. I hope he may turn up yet someday. I have felt miserable since he has been gone, and it is with deep regret that I have to communicate his loss to you . . . .
>
> The weather is very cold, but we stand it very well. Have plenty of clothes. Some shoes wanting. Our boys are in fine health, and our army is in good condition . . . .
>
> From your son,

==========

In his memoirs a young Confederate soldier described the shelling of Sharpsburg (aka, *Antietam*), Maryland, on 16 September 1862, the day before Union and Confederate infantries clashed there:

> Every shell went screaming, whistling, whining over our heads, and not a few burst nearby us. Sometimes shell after shell would burst in quick succession over the village. . . . None of our soldiers were in the town, except the cooks and a few stragglers who hid themselves in the cellars as soon as the bombardment began.

==========

A U.S. Army war correspondent employed by the *Boston Journal* is quoted in *Battles and Leaders, Volume II*:

> I recall a Union soldier lying near the Dunker Church with his face turned upward, and his pocket Bible open upon his breast. I lifted the *[open Bible]* and read the words: "Though I walk through the valley of the shadow of death, I will fear no evil; for thou art with me. Thy rod and thy staff, they comfort me." Upon the *[Bible]* fly-leaf were the *[handwritten]* words: "We hope and pray that you may be permitted by kind Providence, after the war is over, to return."

==========

Born in 1848, a teenage Confederate soldier enlisted in the army in 1861 when he was only 13 years of age. Unaware of his age at that time, the army only knew that he was young and accepted him as a drummer. He was a member of Company K, 8th North Carolina Infantry Regiment. Three years later from Richmond, Virginia, on 1 December 1864 he penned a letter to his mother:

> Dear Mother,
>
> . . . Ma, I witnessed something the other day that I never saw before & never wish to see again. *[A soldier]* of our brigade was shot for desertion. Poor fellow. I could not help pitying him when he was marched around the brigade. We were formed in a hollow square. He was placed upon his coffin, and a white cloth *[was]* placed over his eyes. And then the command was given, Ready, Aim, Fire, and twelve rifles were discharged. Only six were loaded. Five balls struck him, and the doctor dismounted at the discharge and examined his pulse and came back and reported him dead, and we were marched back to our quarters. Oh, it was awful . . . .
>
> Your affectionate son,

==========

A Confederate soldier, who had risen from private to colonel, wrote a letter to his wife on 17 February 1863. He was attached to the 8th Texas Cavalry and wrote the letter at Lewisburg, Tennessee:

> My darling,
>
> The weather was most intensely cold on the entire trip, and more suffering I never saw, I believe. I was told that one man was frozen to death and many were frostbitten. Among the rest, your sweetheart has a pair of frostbitten feet and a finger in the same <u>delightful</u> condition. Don't be alarmed though, for they are nearly well again. . . . I was attacked with paralysis of the arms, spine, and brain which left me senseless and speechless for about 24 hours. They all thought you would be a nice young widow, but God in His mercy has restored me again, and I am now, and have been for several days, in command of my Regiment, doing duty and eating full rations . . . .
>
> Love to Susie and Kit,

==========

A wife in Floyd County, Virginia, wrote a letter to her husband, a Confederate private in the 54th Virginia Artillery Regiment. The letter was written 21 August 1863:

> My dear husband,
>
> . . . I read your kind letter with great pleasure and with a thankful heart to hear that you *[are]* well and hardy, and I do pray to God that He may always add unto us such blessings . . . . I do pray the happy time is not far distant when we may meet again, *[and]* my whole heart's desire is that we may meet soon . . . .
>
> I still remain your true wife till death,

==========

A Confederate officer in a winter camp at Shelbyville, Tennessee, wrote a letter to his sister on 12 January 1863:

> Dear sister,
>
> . . . Thanks to a kind Providence I am again safe. . . . It was pretty cold sleeping Sunday & Monday nights with only one blanket. On Tuesday it was raining pretty much all day and was very disagreeable.
>
> . . . We lay flat down on the wet ground under the most terrific shelling I ever saw. Besides, it was raining, and a groan or scream occasionally would *[come from]* the mangled form of someone who had paid his last sacrifice.
>
> . . . The first of January I commenced to reading the Bible through. I commenced the 1st chapter of Genesis – read two chapters every night . . . . Tonight I read *[the]* 23rd and 24th chapters . . . . I would give anything to see Pa & Ma and you once more. Write as often as you can.
>
> Your affectionate brother,

==========

A Confederate soldier camped at Allisonia (later, *Estill Springs*), Tennessee, wrote to his mother in Louisiana on 18 November 1862:

> Dear Mother,
>
> In my trip through Kentucky the only thing that happened to me was that my shoes gave out, and I had to travel about five hundred miles barefoot. My feet were so

sore I could hardly walk. One of them had thirteen blisters . . . . You can hardly think how the poor men looked marching twenty-five miles a day barefoot, and hardly clothed enough to cover their nakedness. I am not in very good health. I have had the diarrhea for more than two months, and have it yet, the very worst kind. . . . Our company that left home with 112 men is now reduced to 47.

I remain as ever your affectionate son,

==========

A wife in Cornersville, Tennessee, wrote to her husband, a surgeon in the Confederate Army. She penned her letter on New Year's Eve, 31 December 1862:

My dear husband,

Excitement, Suspense, *[and]* Anxiety are stamped on every countenance. All day we have heard the boom of cannon. We have not been told, but *[we]* know all too well from whence it comes. And we know that many brave soldiers now lie cold in death, who only this morning were full of life and hope. . . . I know this is the dark side of the picture, but I cannot banish it from my thoughts . . . . Come home, dearest, as soon as possible. You have no idea how I long to see you again.

Your loving wife,

==========

When the war began Judith W. McGuire lived in Alexandria, Virginia. From her home she could gaze across the Potomac River and see the U.S. Capitol dome. When the U.S. Army occupied undefended Alexandria she fled southward and became a refugee in Richmond. McGuire kept a diary in which she chronicled her volunteer work as a nurse in Richmond's military hospitals. Two years after the war ended her diary, entitled *Diary of a Southern Refugee during the War*, was published. Excerpts are quoted below. Her quoted words begin in early 1861 in Alexandria before she was forced to become a refugee:

4 May 1861: Can it be that our country is to be carried on

> and on to the horrors of civil war? I pray, oh how fervently do I pray, that our Heavenly Father may yet avert it. . . . The embattled hosts of the North will have the whole world from which to draw their supplies; but if, as it seems but too probable, our ports are blockaded, we shall indeed be dependent on our own exertions, and great must those exertions be.
>
> 16 May 1861: Federal troops *[are]* quartered in Baltimore. Poor Maryland! The North has its heel upon her, and how it grinds her! I pray that we may have peaceful secession.
>
> 29 May 1861: I feel so anxious and restless about those so near and dear to us who have gone forth to defend us. The loss of property will be nothing if our boys are spared. I am willing to be poor, but let, oh, let our family circles be unbroken.
>
> 6 June 1861: The ladies assemble daily, by hundreds, at various churches for the purpose of sewing for the soldiers. They are fitting-out company after company. . . . We shall be greatly outnumbered, I know, but *[superior Union]* numbers cannot make up for the zeal and patriotism of our Southern men fighting for home and liberty.

After McGuire reached Richmond she began serving as a nurse in the various military hospitals in the city:

> 7 March 1862: I returned to my post by the bedside of the soldiers. Some of them are very fond of hearing the Bible read. . . . I read of the wars of the Israelites and Philistines, *[and a wounded soldier]* said: "I know why you read that chapter; it is to encourage us, because the Yankee armies are so much bigger than ours.
>
> 10 April 1862: I spent yesterday in the hospital by the bedside of Nathan Newton, our little Alabamian. I closed his eyes last night at ten o'clock after an illness of six weeks. His body, by his own request, will be sent to his mother. Poor little boy! He was but fifteen and should never have left his home.

7 May 1862: While listening to *[a soldier in]* delirium, two gentlemen came in, announcing heavy firing on the river. We had been painfully conscious of the firing before, but remembering that Drewry's Bluff was considered impregnable, I felt much more anxious about the patient than about the enemy.

9 June 1862: Poor fellows! Dearly have they paid with loss of limb and other wounds more painful still. They want to be cured *[so]* that they may be on the field again. "Thank God," said a man with his leg amputated, "that it was not my right arm, for then I could never have fought again; as soon as this stump is well I shall join Stuart's cavalry; I can ride with a wooden leg as well as a real one."

7 November 1862: How it gladdens my heart when I see that a vessel has run the blockade and arrived safely at some Southern port, laden with ammunition, arms, and clothing for the army!

12 February 1863: *[The unidentified soldier's face]* was young, almost boyish, and though disease and death had made sad ravages, they could not conceal delicately carved features, a high fair forehead, and light hair which had been well cared for. . . . Kind gentle hands placed him in his soldier's coffin, and he had a Christian burial at Hollywood *[Cemetery]*, with the lonely word "STRANGER" carved upon his headboard. We trust that his sad story in the *[news]*papers may meet some eye of which he had once been the light, for he was surely Somebody's Darling.

15 April 1863: One *[soldier]* is a youth of seventeen years, who . . . is now sinking with consumption. His case elicits great sympathy and kindness. His name is Stansberry, and he is from Baltimore. We have reason to hope that he is prepared to meet his God.

16 May 1863: The cheerfulness of the wounded, in proportion to their suffering, is much greater than that of the sick. Under my care yesterday was one poor fellow with a ball embedded in his neck, another with an amputated leg,

one with a hole in his breast through which a bullet had passed, another with a shattered arm, and others with slighter wounds; yet all showed indomitable spirit.

28 May 1863: The wounded *[are]* cheerful and doing well. I read, distributed books, and talked with them. . . . I have never but in one instance had an unpleasant word or look from anyone whom I endeavored to treat with kindness. Bible reading is always kindly received.

26 August 1863: The beautiful shade trees and bold spring *[at Jackson Hospital]* are delightful to the convalescents during this warm weather. Fast-day was observed there with great solemnity. I heard a Methodist chaplain preach to several hundred soldiers, and I never saw a more attentive congregation.

8 January 1865: A soldier in the hospital *[asked me]*: "When do you think my wound will be well enough for me to go *[home]*? I want to get married, and the lady don't know that I am wounded, and maybe she'll think I don't want to come *[home]*."

I *[replied]*: "Ah, show her your scars, and if she is a girl worth having she will love you all the better for having bled for your country, and *[recite this poem]* to her:

> It is always the heart
> that is bravest in war,
> That is fondest and truest
> in love.

6 April 1865: *[After our army evacuated Richmond]* I went to the hospital, as was my duty. My dear friend cheered me by being utterly incredulous about the reported surrender. As usual she is cheerfully devoting her powers of mind and body to her hospital. For four years she has never thought of her own comfort, when by sacrificing it she could alleviate a soldier's sorrow.

Undated entry: Enormous wrongs were done to us *[by the North]*, yet we had no desire to do the slightest wrong to even the bitterest of our enemies. We refused not to do

> them justice; we were not unwilling to seek for them the mercy of Heaven, to extend to them the hand of charity, to supply their wants when *[they were]* captured; to attend as far as possible to their sick and dying and dead, and *[we]* asked for nothing from them but that they leave our borders, never to return.

==========

A Confederate private in the Hampton Legion Infantry Regiment wrote a letter to his family on 23 July 1861, two days after the First Battle of Bull Run. He told them about the battle and the many killed and wounded soldiers that his family knew. Then he described the devastation of war:

> Dear Father, Mother, Brother, and Sisters,
>
> . . . . The whole country is laid waste here, the people have left their homes, the fences *[are]* torn down around the corn . . . the people are all gone, and fine gardens of cabbage and Irish potatoes, all laid waste. I thought I knew something about the destruction of war, but *[now I know that]* I never truly knew anything.
>
> Yours truly,

==========

A Confederate private wrote to his wife on 7 July 1863, four days after the Battle of Gettysburg ended. He explained his role in "Picket's Charge" on 3 July. Then he lamented the hardships of life in the Confederate Army:

> My Dear Wife,
>
> . . . We are now cut off from our homes and from all communication with our loved ones there. This is another distressing privation to add to the many other privations and hardships of the soldiers' life. But we must bear this for the good of our Cause. How long! Oh, how long will this cruel war continue!

==========

A Confederate lieutenant wrote to his father on 17 December 1862 and described the Battle of Fredericksburg in Virginia. In that pivotal battle the Union hoped to overrun Confederate defenses at Fredericksburg and then march on to attack Richmond. However, one-at-a-time the Union brigades made ill-advised frontal assaults against Confederates atop Marye's Heights. Confederate infantry was protected by a stone wall and supported by artillery. Union assaults all failed, and the march toward Richmond was abandoned:

> Dear Father,
>
> . . . We have had another great battle at this place resulting in a decisive victory. . . . Small arms made one continual roar without a moment's cessation from 15 before 11 *[o'clock]* until dark. It is said we were attacked by 40,000 federals, whilst our force which opposed them did not amount to one-fourth of that number. . . . We repulsed them 17 different times, each *[Union]* charge being made with overwhelming numbers. . . . The slaughter on the right is said to exceed anything else in the war. It must have been awful. . . . Our army is in fine condition & went into the fight with perfect cheerfulness. . . . Every man stood to his post & fought bravely.

==========

From a camp near Kinston, North Carolina, a Confederate soldier wrote to his mother on 30 April 1863:

> Dear Mother,
>
> I take the present opportunity *[to write]* you to let you know I am well . . . . You said you had given *[up hope of]* ever seeing me again. You must not get out of heart, for I will be home in less than three months. Tell all the children howdy for me.
>
> You wanted to know what I get to eat. I get meat and cornbread. I have not had a biscuit since the box you sent me. We get some crackers sometimes.
>
> Your son until death,

==========

The next week on 5 May 1863 the same soldier wrote to his father. During the week his unit had moved via train to Richmond, Virginia, and had arrived on 3 May:

Dear Father,

We are camped about a mile from town inside the breastworks. We were expecting a fight all day yesterday. The Yankees are eight miles *[away, and there was a fight there]*. Four companies *[from]* our Regiment *[went]* on the train last night to fetch in the sick and wounded. They ran into the Yankees, and they had to turn back. There has been fighting at Fredericksburg. The reports say that our loss is heavy, and the Yankees *[loss is]* still greater.

. . . I like this place . . . but we only get a quarter of a pound of meat a day to *[each]* man and a pound of loaf to *[each]* man. We have just got word that there are twenty-eight regiments of Yankees on the Fredericksburg railroad. They burned up a train of *[our]* boxcars Sunday night.

Write soon,

==========

From Goldsboro, North Carolina, on 14 May 1862 a Confederate private in the 49th North Carolina Infantry Regiment wrote a letter to his sister:

My dear sister,

. . . I have been protected by a higher power when danger was nigh, and I have been thankful for it, *[but]* not as *[thankful]* as I should have been. I have at times been led astray and have wandered from the path of duty. I hope to be humbler, live a better life, and love God more. I pray *[for]* His protection in the hour of danger and his deliverance from disease, death, and the cursing influences of camp life. I pray to live to *[be]* good and be of service to my country. I pray to get back home to comfort my poor mother and make her happy . . . .

Write soon,

==========

A lieutenant colonel in the Confederate Army commanded the cavalry element of the famous Hampton Legion. On 5 January 1862 he wrote to his wife in Edgefield District, South Carolina. The letter was written at Camp Wigfall in Virginia:

> My Darling Wife,
>
> . . . It is now exceedingly cold, but I don't suffer from the cold. A good many of our men have been *[ice]*-skating *[on the frozen river]* for the last day or two. One poor fellow from the Georgia regiment was drowned yesterday. Two men were skating when the ice broke, and they both went down. This Georgian jumped in and saved them – and afterwards went back to show *[others]* how he saved them, when the ice broke. He went down and drowned before they could get him out.

The same lieutenant colonel again wrote to his wife the following month on 2 February 1862:

> My Dearest Wife,
>
> This is the rainiest – the snowiest – muddiest, the most disagreeable country. . . . . Now it is raining. Night before last it snowed. Yesterday it thawed, and it seems that everything combines to keep the earth saturated with water.

The officer wrote again on 26 February 1862:

> My Darling Leila,
>
> . . . We cannot see, my darling, into the future, but I trust & have confidence in our people to believe that if the unprincipled North shall persist in her policy of subjugating the South, that we who are able to resist them will continue to do so until we grow old and worn out in the service, and that then our sons will take the arms from our hands and spend their lives, if necessary, in battling for Liberty and Independence.
>
> . . . Have you ever anticipated, my darling, what would be our probable *[fate]* if we should be conquered in this war? The picture is too horrible to contemplate. . . . We would be the most wretched and abject people on the face of the earth.

==========

An Alabama native enlisted in Company H of the 8th Louisiana Infantry Regiment. The young Confederate private wrote a letter to his sister on 28 May 1862. At that time his regiment was camped in Madison County, Virginia:

> Sister Mary,
>
> . . . Home. Home. How much pleasure is in that word, home? There is more than a tongue can express. How oft have I thought of home, that place that I formerly so little appreciated. And to think of those who are there – the kind father, the indulgent mother to *[whom]* I have been so disrespectful in days gone by. The fond sisters that I have so oft mistreated. Oh, that I could have my time over again, how different I would live.
>
> . . . I will stay in the *[army]* forever before I will allow my country to be invaded. I will submit to the toils and hardships of camp. . . . I will endure the toil *[and]* forebear the pain produced thereby, before thinking of submitting to such Yankee vandals as those thieving, undermining, careless, unprincipled band of demons, which are really beneath the notice of the Devil himself . . . .
>
> Your warrior brother until death,

The soldier did remain in the army. In the spring of 1864 he was wounded, but he survived and was hospitalized for two months. After his release from the hospital he was granted a wound furlough and put on a train bound for home, but he died before he got there. His friends prepared the following obituary:

> OBITUARY
>
> Died at the "Wayside Home," Augusta, Georgia, on the 6th of April 1864, Thomas D. Newton, of Co. H., 8th Louisiana Regiment, in the 24th year of his age. He had remained nearly two months in a hospital at Lynchburg, Va., and . . . having partly recovered, had obtained a furlough and was on his way home, when he had a relapse and died.
>
> . . . It is thus that another . . . sacrifice has been offered on the altar of our country. Among the brave, he was the bravest; fearing nothing in defense of his country, in whose service he lost his life with hope that she might one day be free. From his early youth he was deeply impressed with a love of freedom, his country, and his home. . . . The first sound of the drum, calling our men to arms, *[had]* aroused

> him, and he went with the motto, "Conquer or Die." With his immortal "Stonewall" he, often cold, hungry, and fatigued; traversed the mountains and valleys of Virginia; invaded Maryland and Pennsylvania; and participated in every battle fought by that army . . . .
>
> . . . But oh! He is gone, and gone forever! Wail, Alabama, wail!

==========

The month after the war ended a Union private from Company A, 36th Iowa Volunteers, wrote to a female friend. The letter was dated 24 May 1865 and mailed from Saint Charles, Arkansas:

> Miss,
>
> . . . You stated in your letter that many changes had taken place up in Monroe, *[Iowa]*, here of late. The most heart-breaking of all is my darling has long ago forsaken me and married *[someone else]* and left me here to mourn my life away . . . . I thank you very much for your sympathy toward me in my sad bereavement.

==========

Several days after the pivotal Battle of Gettysburg on 1-3 July 1863 a Confederate private wrote a letter to his brother. The brother had been wounded at Chancellorsville two months earlier on 3 May and was a patient in a military hospital in Augusta, Georgia:

> Dear Brother,
>
> Gettysburg!!! 'Twas an awful place, four days of volleyed thunder, the worst you ever heard tell of . . . .

==========

A Union corporal was a prisoner of war in Lawton POW camp near Millen, Georgia. On 14 November 1864 he wrote a letter to his wife in Essex County, New York. He had not gotten a letter from her or from his friends during the past seven months:

> Dear Wife,
>
> I have written to you every month since I was captured the 5th of May *[1864]*, and have seen hard times since

> *[then]*. I have been a prisoner for 6 months and 9 days. . . . I have not heard from any of my friends since the last letter I got from you *[in April before I was captured]*, but I hope you and little Charlie are well.
>
> If you want any money write to Martin if he is alive, for *[he owes me]* nine dollars and a half. That will help you some. Do what you think is best as it regards our affairs. Perhaps you had better sell the cow if hay is dear. . . . There is no use in trying to get a letter to me, for it will never be delivered to me if it should come here.
>
> Yours affectionately,

==========

On 21 November 1864, Abraham Lincoln, President of the United States, wrote a letter to Mrs. Lydia Bixbey. She was the mother of five sons, all of whom had joined the U.S. Army. All five young brothers had been killed in battle during the war.

This handwritten letter to Mrs. Bixbey is the most famous letter ever penned by Lincoln. It was read, verbatim and in its entirety, 134 years later in 1998 in the motion picture, *Saving Private Ryan*:

> Dear Madam:
>
> I have been shown in the files of the War Department a statement of the Adjutant General of Massachusetts *[which states]* that you are the mother of five sons who have died gloriously on the field of battle.
>
> I feel how weak and fruitless must be any words of mine which should attempt to beguile you from the grief of a loss so overwhelming. But I cannot refrain from tendering you the consolation that may be found in the thanks of the Republic they died to save.
>
> I pray that our Heavenly Father may assuage the anguish of your bereavement, and leave you only the cherished memory of the loved and lost, and the solemn pride that must be yours to have laid so costly a sacrifice upon the altar of freedom.

Confederate letter postmarked from Georgetown, South Carolina.

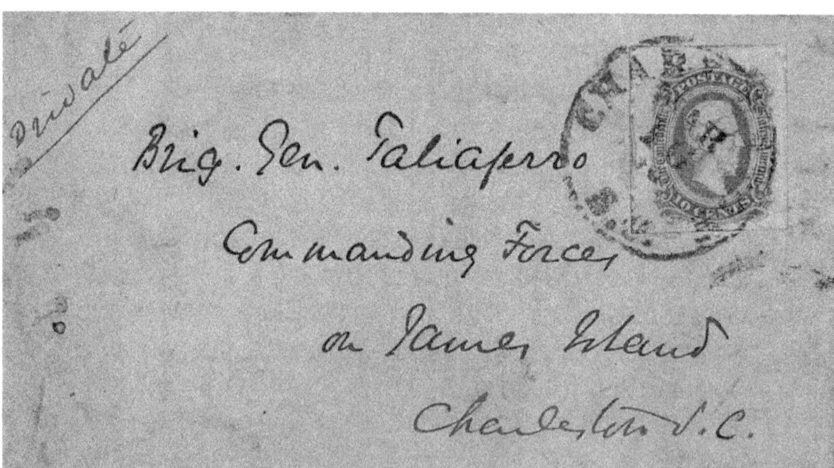

Confederate letter postmarked from Charleston, South Carolina.

The Southern Icon

> My religious belief teaches me to feel as safe in battle as in bed. God has fixed the time of my death. I do not concern myself with that.
> *[Gen. Thomas J. "Stonewall" Jackson, quoted posthumously in Stonewall Jackson, 1898]*

Overview: Gen. Thomas J. "Stonewall" Jackson (1824-1863) became a warrior like no other, the Achilles of the Nineteenth Century. He embodied all traits of a successful military leader: audacity, tenacity, mobility, discipline, fearlessness, knowledge of the enemy, and an obsession with offensive warfare. Yet, Jackson's greatest strength lay in his uncanny ability to inspire his soldiers.

Eccentric and obsessive, Jackson viewed the war as a spiritual crusade. Fanatically religious, he lived the role of an obedient Old Testament warrior. Like the biblical Joshua armed with the wrath of God, he led his men into battle to slay his enemies. He was a devout Christian soldier in every sense of the word. His fame flashed across his beloved Confederacy, soared over the land of his enemies, and dazzled European governments.

On 2 May 1863, Jackson led his soldiers on a secret foray around the flank of a Union army over three times his size. He achieved total surprise and routed the entire Union force. That night Jackson rode far ahead of his troops to reconnoiter enemy positions. In the dark in the most tragic *friendly fire* error in American history, his own soldiers shot him three times. He died eight days later.

Early Life and Hardships: As a youth, Thomas Jackson's life was marked by misfortune and privation. He was born on a hardscrabble farm near Clarksburg, Virginia (later, *West Virginia*), on 21 January 1824. When he was two years of age his father and his older sister died of typhoid fever. His widowed mother, Julia Jackson, quickly found herself deeply in debt. She moved into a tiny two-room cabin, taught school off and on, and took in sewing in order to put food on her table.

Thomas Jackson was six years old when his mother remarried in

1830. She died the following year and was buried in a homemade wooden coffin in an unmarked grave in Westlake Cemetery. Thomas, now a full orphan, disliked his stepfather, and the sentiment was mutual. The next year Thomas, age seven, left home and lived with his half-uncle, Cummins Jackson. Four years later at age 11 he left his half-uncle and lived with his aunt, Polly Brake.

Polly's husband and young Jackson, self reliant but shy, failed to get along well. Jackson endured continual verbal abuse. After a year he ran away, walked 16 miles through a mountain wilderness and found his way back to Cummins Jackson. Cummins sheltered the young runaway and put him to work on the family farm.

> You may be whatever you resolve to be.
> *[Gen. Thomas Jackson, List of Maxims, 1863]*

Now a teenager, Thomas would remain with Cummins for seven years. He learned to use a sheepdog and began tending sheep. He learned to handle oxen-drawn wagons, and he helped with planting, cultivating, and harvesting corn, wheat, and vegetables. In those days, schooling was not free. When not laboring for his half-uncle, Jackson saved the money he made while working for others on the side. He paid tuition and attended school off and on. He learned only the basics of reading, writing, and elementary mathematics, and that was the extent of his academic training.

Jackson knew the importance of furthering his meager education. With no money to pay for more schooling, he decided to teach himself. He could read elementary text, and he theorized that the more he read the better he would become at it. However, he had a problem. Farm work was a dawn to dusk endeavor, so Jackson needed light to read at night. He struck a deal with his uncle's slave. Jackson would teach the slave to read, and in return the slave would keep Jackson supplied with flammable pine-knots. Teaching a slave to read violated Virginia law, but the deal worked. With Jackson as a tutor, the slave learned to read basic text. Blazing pine-knots provided enough light so that Jackson and the slave could practice reading borrowed books each night.

West Point and the Mexican War: At age 18, Jackson searched for a low cost means of attending school. He applied for an appointment to the U.S. Military Academy at West Point, but he was rejected. Fate then stepped in. The congressional appointee dropped out after his first day at West Point, and Jackson replaced him.

Shy and socially awkward, Jackson struggled mightily at West Point. He did not get along well with culturally polished sons of rich and influential families. Further, although his self-education efforts through reading at night had helped him, he was woefully ill prepared for the rigid academic regimen at West Point. Although he tried hard, he almost got dismissed for academic failure. However, his dogged grit and determination plus his extraordinary memory and powers of concentration swayed the military faculty. Jackson remained at the academy. Each year he moved up in cadet academic rankings. When he graduated in 1846 he ranked 17th in his class of 59 cadets.

Upon graduation the U.S. Army commissioned Jackson as a second lieutenant. The army sent the young officer to join American forces during the Mexican War (1846-1848). Thomas Jackson found his calling. With the First U.S. Artillery Regiment he served with distinction during battles at Veracruz, Contreras, Chapultepec, and Mexico City. At Chapultepec he violated a direct order by not withdrawing his artillery under fire, but an investigation proved that his decision was proper. Jackson's audacity and initiative during the war earned him three brevet promotions.

> Sacrifice your life, rather than your word.
> *[Gen. Thomas Jackson, List of Maxims, 1863]*

After the war the army dispatched Maj. Jackson to army forts in New York, and then to forts in Florida during the Seminole Wars. At Fort Meade, 25 miles south of Tampa, Jackson rose to second-in-command. During this time he began to embrace Christianity. His spiritual faith soon would become the bedrock of his character. He felt his military experience would serve him well in an instructing capacity, so he began searching for a military teaching position.

Ten Years at Virginia Military Institute: In early 1851, Jackson left active duty and accepted a professorship at Virginia Military Institute (VMI) in Lexington, Virginia. VMI was one of the two premier military academies in the United States. Jackson became Professor of Natural and Experimental Philosophy (astronomy, physics, acoustics, and various other sciences) and Instructor of Artillery. The new professor soon became unpopular with his cadet students. He would study long into each night and memorize, word for word, his lectures. In class the next day he would recite the lecture, from memory, to cadet students. Any cadet who asked for after-class help would be given the same lecture again. Few cadets asked for help a second time.

Maj. Jackson's eccentric habits, quirks, and his cold demeanor contributed to his unpopularity. He believed that sitting in a chair put undue stress on his internal organs, so he preferred to either stand erect or lie flat on his back. In class he rarely sat down, and when lecturing he always stood. Jackson thought his left arm was slightly longer than his right. Consequently, while lecturing he would hold his left arm overhead with the palm facing forward. He believed that this tactic would eventually equalize blood circulation and shorten his arm. Instead of conventional tests or examinations, Jackson often required cadets to stand in front of the class and recite the essence of his lectures. Behind his back, cadets made fun of their professor and called him Tom Fool Jackson (some historians later would theorize that Jackson was affected by Asperger Syndrome).

> Speak but what may benefit others or yourself, and avoid trifling conversation.
> *[Gen. Thomas Jackson, List of Maxims, 1863]*

While he was at VMI, Jackson married Elinor Junkin in 1853. The marital union was happy but brief. The year after their marriage Elinor gave birth to a son, but he was stillborn. Elinor hemorrhaged and died within the hour. The widower married Anna Morrison in 1857, and Anna gave birth to a daughter, Mary, the next year. The infant died in less than a month (four years later during the war, Anna and Thomas would have another child, Julia). Years later Anna's memoirs would portray the marriage as both passionate and tender. Further, at home Jackson was completely informal, even playful, in contrast to his public persona.

Maj. Jackson remained unpopular with his cadet students. In addition he refused to engage in social contacts with contemporaries in Lexington, exclusive of members of his beloved Presbyterian Church. Jackson's deep religious beliefs had grown to become the center of his life, and he evolved into a man of deep piety. He believed that the hand of God was at work everywhere and that he was a mortal instrument of God's will.

At church during 1856 he organized and taught an afternoon Sunday School class for free Negro and slave members. He believed they would profit from what he called "the strong hand of the Gospel." In class he would call each student by name, and in return they would address him as Marse Major. On two occasions slave women implored Jackson to buy them. They knew him to be fair and kind, and they believed he would let them earn their way to freedom. The church pastor, Dr. William White, wrote of the close relationship between Jackson and his Negro students:

> In their religious instruction he succeeded wonderfully. His discipline was systematic and firm, but very kind. His *[students]* revered him and loved him, as they would have loved a brother or father.

A deranged abolitionist, John Brown, raided and occupied a federal arsenal at Harpers Ferry, Virginia, in October 1859. He was captured, convicted of treason, and sentenced to death. The Virginia governor sent a large contingent from the VMI cadet corps to Harpers Ferry for security purposes. Maj. Jackson commanded the cadet artillery at the public hanging. In death John Brown evolved into an abolitionist martyr, and no soothsayer was necessary to perceive that problems between North and South loomed on the horizon.

American Civil War – 1861: Virginia seceded from the United States on 17 April 1861 and joined the new Confederate States of America. Maj. Jackson became the drill-master for C.S. Army troops in the Richmond area. He drilled new soldiers relentlessly, believing that ironclad discipline was the key to battlefield success. Quickly promoted to the rank of colonel by the governor, Col. Jackson took command of the 2nd, 4th, 5th, 27th, and 33rd Virginia Infantry Regiments. He marched his soldiers into the Shenandoah Valley, the breadbasket of the Confederacy, to defend western Virginia from Union attack. Col. Jackson surprised the North with successful raids against the B&O Railroad. His tenacity and initiative earned him a promotion to brigadier general in June 1861.

The next month on 21 July 1861, Gen. Jackson and his brigade made their immortal defensive stand on Henry Hill at the Battle of First Bull Run (aka, *Battle of First Manassas*). During the fierce struggle Jackson mimicked his classroom habit. He held his left arm high overhead, palm facing forward. Some who did not know him believed the unnatural act was an entreaty to God for victory. Others saw the act as a "stop" signal for attacking Union troops.

Jackson and his raised arm were highly visible, and he became a perfect target for northern soldiers. Yet, he was wounded only once, a Minie-ball through his raised left hand. It would prove to be a clean wound, so amputation was not necessary. Jackson's brigade suffered more casualties than any other C.S. Army brigade that day. Yet, the brigade held its ground until reinforcements arrived. The tide of battle turned, northern troops fled in disarray, and the Confederacy won the first major land battle.

> My preservation was entirely due, as was the victory, to our God, to whom be all the honor, praise, and glory.
> *[Gen. Thomas Jackson, in a letter to his wife the day after the battle, 22 July 1861]*

Jackson's fame skyrocketed after the Bull Run victory, due in part to his new *nom de guerre*, "Stonewall" Jackson. Thereafter his brigade would be called the Stonewall Brigade. Two months later the Confederacy promoted him to the rank of major general and gave him command of the entire Shenandoah Valley District. In camp and in battle Jackson rode a former Union horse, a gelding that he named Little Sorrel.

American Civil War – 1862: When battle season began in the spring of 1862 it seemed that Gen. Jackson and his soldiers were invincible. Jackson had slightly less than 17,000 men in his command. Vastly outnumbered by northern troops, he wisely never engaged in set-piece battles. Instead he attacked his enemy's flanks, his rear, his supply lines. Gen. Nathaniel Banks, the Union commander, tried in vain to trap Jackson's army. However, Jackson could not be cornered. It seemed that he was nowhere. Yet, he was attacking everywhere. In the brutal summer of 1862 one historian explained:

> Jackson won significant victories with a force of about 17,000 against a combined northern force of over 60,000. *[That summer]* Jackson became the most celebrated solider in the Confederacy.

In addition to major victories, Jackson and his soldiers triumphed in raids and surprise attacks. Jackson had become a consummate military commander: aggressive, audacious, highly mobile, and able to inspire his men to the point of near fanaticism. Because Jackson and his force moved so rapidly and so often, his soldiers began calling themselves Jackson's Foot Cavalry. To northern troops, Jackson and his army were a lethal will-o-the-wisp.

> Oh! Divine art of subtlety and secrecy!
> *[Sun Tzu, The Art of War, circa 500 BC]*

The southern public got war news via newspapers that printed exuberant accounts of Jackson's successes. These victories boosted morale throughout the Confederacy. To the average citizen, Jackson became larger than life, a living legend, a hero in every sense of the word. In warfare he had an intuitive instinct to take advantage of every tactical opportunity. However, on Sunday he never engaged in any secular conversation or any form of non-military work. Also he never read newspapers or wrote letters on Sunday. He often sucked on lemons during a battle, believing the bitter fruit would stimulate his mental faculties.

These personal eccentricities and his rigid Presbyterian faith only added to Jackson's luster and mystique. By the late spring of 1862, Jackson had evolved into the most famous military commander in the world. Newspapers throughout the North and South chronicled his eccentricities and military successes. Editors began calling Jackson the "American Napoleon."

In the early summer of 1862 the Union Army marched toward Richmond. Jackson and his brigade left the Shenandoah Valley and joined Gen. Robert E. Lee, commander of the Army of Northern Virginia. In what became known as the Seven Days Battles, Jackson operated under Lee's command. Gen. Jackson and his soldiers were heavily engaged at Gaines Mill, Savages Station, White Oak Swamp, and Malvern Hill. These set-piece and frontal assault battles afforded

Jackson few opportunities to use surprise and mobility as he had done in the past. Although combined southern forces repulsed the Union advance, Jackson's famed unpredictability and brilliant tactics were not on display.

> Jackson's genius never shone when he was under the command of another.
> *[Gen. Daniel Hill, in a letter, summer 1862]*

Gen. Lee reorganized the C.S. Army into two wings. Gen. James Longstreet commanded the Right Wing. Jackson took command of the Left Wing, and his brilliant tactics again would be showcased. In late August 1862 he marched his loyal "foot cavalry" all the way around the flank of his enemy and got behind him. At Manassas Junction he captured the enormous Union Army supply depot. He and his men then feigned retreat to entice the Union commander, Gen. John Pope, to attack. Pope obliged. Big mistake!

Joined by Gen. Longstreet and his Right Wing, Jackson and his soldiers thrashed Union forces on 30 August. The battle was fought on the same ground as the decisive Confederate victory 13 months earlier, and the result was the same. Gen. Pope and his army fled toward safety in Washington. Southern troops scoured the battlefield and captured weapons, munitions, and prisoners. Thus ended the Battle of Second Bull Run (aka, *Battle of Second Manassas*).

> In war there is no second place prize for the runner-up.
> *[Gen. Omar Bradley, in Military Review, 1950]*

Bad news sells newspapers. As casualty reports grew, northern newspaper editors became ever more critical of continuing the war. In the North, public support for the war waned. Aware of the war's unpopularity in northern states, Gen. Lee sought a decisive victory that would force northern politicians and citizens to recognize the Confederacy as a sovereign nation and end the fighting. To gain such a victory he decided that he must invade the North.

Circling around Washington, Lee crossed the Potomac River and led his army into Maryland. Jackson captured Harpers Ferry and then drove northward to join Gen. Lee near the little community of Sharpsburg. There on 17 September 1862 the C.S. Army and U.S. Army clashed in what became the bloodiest single day in American history. That day America would suffer four times as many casualties as on D-Day in Europe 82 years later.

The deadly Battle of Sharpsburg (aka, *Battle of Antietam*) pitted the South against the numerically superior forces of Gen. George McClellan, the northern commander. The South dug in and waited. The North would attack, for the South had invaded their country.

Jackson and his Left Wing soldiers bore the brunt of repeated federal attacks in the morning. Against overwhelming odds his men held their ground. Late that afternoon they helped repulse frontal assaults against the Confederate center. An historian would note:

> Confederate forces held their position. McClellan failed to drive the South out of Maryland. Yet, on the other hand, McClellan had not been defeated either, so Lee had not won the decisive victory that he sought.

The horrible slugfest ended in a draw. Neither side had earned a victory. However, Lee and his army were almost out of munitions. With no means of resupply he was compelled to return to Virginia to rest his army and regroup. In recognition of his stellar performance at Sharpsburg, Jackson received a promotion to lieutenant general on 10 October. His command was redesignated as Second Corps.

On 23 November, Jackson got a most welcome letter. His wife, Anna, had given birth to a healthy baby girl named Julia. Jackson soon would get another gift. Gen. "Jeb" Stuart, the flamboyant and nattily dressed cavalry leader, had noted that Jackson's uniform was torn, colorless, and threadbare. From a tailor in Richmond, Stuart ordered a fancy new uniform coat for Jackson and presented it to him as a present. The eccentric Presbyterian warrior was embarrassed. He claimed that the new coat, although appreciated, was much too handsome for him. He said he would not wear it, but that he would keep it as a treasured war souvenir. His command staff insisted that he wear it to supper at least one time, and Jackson complied. His staff then secretly invited all staff officers to gather and witness their lavishly dressed commanding general.

> Jackson was fanatical in his Presbyterian faith, and it energized his military thought and character. Theology was the only subject he genuinely enjoyed discussing. His

> dispatches invariably credited an ever-kind Providence.
> *[Stephen W. Sears (1931 —), civil war historian]*

In December the Union Army made another attempt to crush the rebel army and capture Richmond. After fighting in Fredericksburg, the South defended a fortified ridge outside the city on 13 December 1862. Seven Union divisions – one brigade at a time – would charge *uphill* across an enormous open field toward ranks of Confederate defenders shielded behind a stone wall and supported by artillery. LtCol. Edward Alexander, a Confederate artillery commander, was standing ready with cannister, and he confidently remarked:

> A chicken could not survive on that field when we open *[fire]* on it.

Union generals ignored one of the most basic tenets of offensive warfare. They sent their brigades *uphill* over open terrain toward an entrenched enemy atop the hill. Perhaps they had never read Sun Tzu's adage: "It is a military axiom not to advance uphill against the enemy." One by one the northern brigades charged. One by one they were mowed down.

Insanity reigned all afternoon. Lee, Jackson, and Longstreet won a crushing and lopsided victory. The vast sloping field below the protective stone wall had turned into a charnel house. Union corpses and thousands of writhing wounded soldiers covered the ground. Looking at the carnage, Gen. Lee told his two corps commanders:

> It is well that war is so terrible, for *[otherwise]* we would grow too fond of it.

The Confederacy experienced wild jubilation over the victory. The *Charleston Mercury* newspaper would editorialize: "Our army has yet known no such word as fail." The *Richmond Examiner* would extol the "stunning defeat to the invader, a splendid victory *[for]* the defenders of our native soil."

News of the horrible defeat devastated the North. Andrew Curtin, governor of Pennsylvania, rushed to the White House to explain the loss to President Lincoln. Curtin told the president:

> It was not a battle. It was a butchery.

American Civil War – 1863: When 1863 began the U.S. President appointed Gen. Joseph Hooker to lead the Army of the Potomac. During two years of war there had been four major Union attempts to capture Richmond and destroy the Confederate Army. All of them had failed. With a new commander Lincoln would try for the fifth time, and his chance of success appeared excellent.

Gen. Hooker planned to use infantry, cavalry, and artillery in a bold double envelopment of the rebel army. He had huge resources at his disposal including 133,800 troops and 413 artillery pieces. In contrast the South had only 59,300 men and 216 guns. Better yet, Hooker reasoned, many of the Confederate units were scattered over the state of Virginia in defensive positions. Much of the South's First Corps was defending near Norfolk, and two divisions were foraging 130 miles south of Richmond. Hooker's well provisioned army would outnumber the poorly supplied South almost three to one.

> Moral forces, rather than numbers, decide victory.
> *[Napoleon Bonaparte, Maxims of War, 1831]*

Hooker crossed the Rappahannock River, and his infantry massed near a small Virginia plantation called Chancellorsville. When the rebel army marched out to defend Richmond, Hooker would have an ace hidden up his sleeve. He had 40,000 Union soldiers waiting near Fredericksburg, to the east. They would march west and strike the Confederate Army from the rear – a perfect double envelopment.

For the South the tactical situation looked bleak. The primary threat was Hooker, entrenched near Chancellorsville. However, Gen. Lee could not ignore the threat to his rear. Consequently he had to divide his already badly outnumbered army. He sent 11,000 troops toward Fredericksburg to block the Union threat lurking there. He would have sent more, but he dared not spare them.

The night before battle Gen. Lee and Gen. Jackson conferred. At Chancellorsville they knew that they could not repel a frontal assault by Hooker's legions without horrible and unacceptable casualties. They devised a risky plan. Once again they would divide what was left of their army. Lee would retain only 13,000 men to face 70,000 waiting soldiers of Gen. Hooker's army. Unthinkable! Gen. Jackson would take the other 21,000 Confederates on a flanking march and attack Hooker's right flank or his rear.

Battle of Chancellorsville – Day One: Led by Jackson riding on Little Sorrel, the flanking march began early on 2 May 1863 at 7:00 a.m. In the distance Jackson could see the Union observation balloon *Eagle* high overhead. Hooker got word of the Confederate troop movement, but he assumed the rebels were retreating. He took no action to interdict them.

Jackson, wearing a black raincoat and with his treasured copy of Napoleon's *Maxims of War* in his haversack, circled 12 miles toward the Union right flank. The ploy worked to perfection. Jackson and his cavalry commander, Gen. Fitzhugh Lee, rode far ahead. Late in the afternoon at 5:15 p.m. they spied the unguarded right flank of the Union Army. They knew they could get behind it. They watched Union troops, who were not expecting hostilities until the next day or two. Northern soldiers had bivouacked for the night. After they had unloaded and stacked their weapons they began napping and cooking rations for supper.

> Jackson's eyes burned with a brilliant glow. . . . His lips were moving. I knew what he was doing. Stonewall Jackson was praying in full view, and in the rear, of general Hooker's right flank.
> *[Gen. Fitzhugh Lee, speaking to the Association of the Army of Northern Virginia 16 years later in 1879]*

Jackson rode back to his corps and organized his divisions into assault waves over a mile long. He was surrounded by his staff, who were his former VMI students. Now they were his officers and his comrades-in-arms. He told them: "The Institution will be heard from today." Silently the grey-clad soldiers of Second Corps slipped through thick woods toward the unsuspecting Union rear. They crept to within 70 yards of the federal troops. One Confederate soldier later would write:

> The Yankees had unloaded and stacked their arms. They were cooking rations, eating, and playing games in carefree fashion. They were completely unaware that Stonewall Jackson's entire corps was about to pounce on them.

Startled deer and rabbits ran out of the woods and into the Union camp, the first indicator of the coming firestorm. The South's first wave began shrieking like demons, and then they charged.

Panic swept through Union ranks, and a total rout began. "Press on! Press on! Press on!" shouted Jackson, riding close behind his first wave. Blue-clad soldiers who were not killed or captured fled from the onslaught. Confederate waves rolled forward, ever forward, until nightfall halted their advance.

Jackson and his staff rode ahead to reconnoiter Union positions and evaluate the practicality of a night attack under the light of the full moon. As they tried to return in the dark they startled jittery pickets from the 18th North Carolina Infantry Regiment. Who was riding toward them? Was this an enemy trick, a ruse? One of the pickets shouted: "Halt, who goes there?"

"Fire!" yelled Maj. John Barry. Confederate soldiers unleashed multiple volleys into the night toward the unseen riders. Three bullets hit Jackson, two in the left arm and one in the right hand. Darkness, tangled vegetation, and the fog of battle prevented prompt medical aid. Eventually a stretcher was brought for the general. In the dark he was dropped twice before reaching a forward aid station.

Later that night at a field hospital an army surgeon, Dr. Hunter McGuire, realized that Gen. Jackson's left arm could not be saved. McGuire amputated it at the shoulder, and the anaesthetized general would recover at the field hospital for two nights.

Battle of Chancellorsville – Day Two: Early on 3 May remaining Union forces were battered and reeling. By mid-morning Jackson's corps and Lee's soldiers linked up, and together they trounced the unnerved Union Army. Meanwhile at nearby Fredericksburg the 40,000 Union troops retreated back across the Rappahannock.

> These were intoxicating times in which *[Jackson]* had risen from an eccentric professor at Virginia Military Institute to the war's most celebrated general. *[Chancellorsville]* elevated the already famous Jackson to mythical status.
> *[Michael E. Ruane, in The Washington Post, 29 April 2013]*

The Confederate victory was complete. The audacious gamble resulted in the greatest coup of the war. Chancellorsville became the undisputed zenith of Confederate military prowess, the highwater mark of the War for Southern Independence. Against all odds the outnumbered ill-supplied South had defeated the Union juggernaut. Little *David* had slain mighty *Goliath*. The improbable Confederate victory at Chancellorsville went into history books as the greatest triumph of Nineteenth Century warfare. One historian noted:

> Despite being outnumbered by a ratio of almost three to one, the *[South]* won arguably its greatest victory of the war, sometimes described as the "perfect battle."

The Confederate victory both humiliated and shocked the North. All odds had favored the massive and well-supplied Army of the Potomac. How could they have lost? The victory brought the South to the brink of independence, and it elevated already iconic Jackson to mythical status. He had risen from an eccentric professor to become the most celebrated military general in America. Hearing the bitter news, President Lincoln collapsed at his desk. He buried his face in his hands and lamented:

> My God! My God! What will the country say?

Death at Guinea Station: On 4 May on a rainy night a two-horse ambulance, escorted by cavalry, stopped at the plantation of Thomas Chandler near Guinea Station, Virginia. Inside the ambulance lay Gen. Jackson. Dr. McGuire, who had stayed with the general, moved him onto a bed in the small plantation office building. If Jackson progressed satisfactorily, in several days he would be carried to his home to complete his recovery.

Jackson rallied. Within two days his facial color returned, and he freely conversed with the surgeon and members of his staff. All signs indicated that he was well on the road to recovery. He told Dr. McGuire that he agreed with the loss of his arm and that God had ordained the amputation. Gen. Lee, informed of the situation, sent Jackson a letter and lamented in part:

> Could I have directed events, I would have chosen – for the good of the country – to be disabled in your stead.

However, on the morning of 7 May, Jackson was nauseated. He had extreme pains in his chest. He was feverish, and he asked for a moist towel. Dr. McGuire realized that these symptoms indicated pneumonia. He applied mustard plasters and administered opiate painkillers. He even tried the old practice of cupping. In the days before antibiotics there was little else that he could do.

> The disease rapidly ran its course. Jackson slipped in and out of delirium. His wife, Anna, was summoned and arrived with Julia, their infant daughter. *[Jackson]* grew worse Friday and Saturday *[8 and 9 May]*.
> *[Michael E. Ruane, in The Washington Post, 29 April 2013]*

Early on 10 May 1863, Rev. Tucker Lacy prayed with Jackson, and Anna sang hymns for him. Lacy told the general that the entire Confederate Army was praying for him. "Thank God, they are very kind," Jackson replied. Then he asked what day of the week it was. Lacy told him that it was Sunday.

> It is the Lord's day. My wish has been fulfilled. I have always desired to die on a Sunday.
> *[Gen. Thomas Jackson, speaking on his deathbed, 10 May 1863]*

Gen. Lee had been informed that Jackson had pneumonia. He dispatched a note to Rev. Lacy. The handwritten missive read:

> Give General Jackson my affectionate regards, and say to him: he has lost his left arm, but I *[have lost]* my right.

By midday a large crowd had gathered outside. They stood in silence. Inside, Jackson was slipping away. His mind wandered, and in delirium he began shouting battle orders: "Order A.P. Hill to prepare for action! Pass the infantry to the front! Press on! Press on!" Dr. McGuire scribbled down a written account of the general's final moments. He noted that, very suddenly, a smile of ineffable sweetness spread over the general's pale face. Jackson spoke quietly, yet firmly and distinctly as if in relief:

> Let us cross over the river and rest under the shade of the trees.

Jackson then breathed his last and died at 3:15 p.m.

Aftermath: Jackson's death would prove to be the greatest personal loss ever suffered by the Confederacy. His body lay in state in the governor's mansion in Richmond. VMI issued a General Order with regard to his death:

> It is the painful duty of the Superintendent to announce to the officers and cadets of this institution the death of their late associate and professor, Lieut. General Thomas J. Jackson . . . .

The funeral on 15 May was one of the largest public gatherings in the South. After the funeral Jackson's body was taken to his adopted hometown, Lexington, and buried in what would become Stonewall Jackson Memorial Cemetery.

> All that was mortal of our great and good chief, Gen. T.J. Jackson, was consigned to the tomb on Friday last. . . . He has been brought back to sleep among us – a world renowned Christian Hero.
> *[Lexington Gazette newspaper, 20 May 1863]*

Throughout the South the mystical hero's death dealt a crushing blow to morale. The loss crippled the Confederacy. Gen. Jackson's martial and stern attitude, his evangelical zeal, and his successes in battle had made him an invincible leader, a military and spiritual champion, an icon, a guarantor of southern independence and liberty. His fame, spread by newspaper reports, had dazzled the South and demoralized the North. Yet, now he was gone.

In the North, news of Jackson's death on 10 May did not reach the press until two days later. That was nine days after the disastrous Union defeat at Chancellorsville. During those nine days most northern newspapers had been replete with gloom and doom. The well-equipped Union army had vastly outnumbered the South. Confederate forces had been undermanned and poorly equipped, and many rebel soldiers had been barefooted. How could Jackson's men, seemingly hopelessly outnumbered, have routed Hooker's legions? Newspaper editors demanded to know when this madness would end. Many began lobbying for recognition of the Confederate States as a sovereign nation and an end to the brutal war.

> Four hostile newspapers are more to be feared than four thousand enemy bayonets.
> *[Napoleon Bonaparte, Maxims of War, 1831]*

Northern editors had been publishing scintillating accounts of northern military debacles. The defeat at Chancellorsville provided yet more fuel for the fire, and Jackson had been the chief instrument of the Union loss. Northern citizens and politicians increasingly had begun demanding an end to the conflict.

> General "Stonewall" Jackson was badly wounded in the arm at the Battle of Chancellorsville and had his arm amputated. The operation did not succeed, and with pneumonia setting in, he died on the 10th inst. near Richmond, Virginia.
> *[Harpers Weekly magazine, 23 May 1863]*

News of Jackson's death swayed public opinion. The southern champion had fallen. Without the fanatical Jackson, the southern Achilles, perhaps the South could be militarily defeated. Northern clamor for peace-at-any-price began to moderate, and the war would drag on for another two years.

If Jackson Had Survived . . . : Many historians have theorized how events might have unfolded if Jackson had survived his amputation. Would his recovery have altered the war's outcome?

After the unexpected Union debacle at Chancellorsville, northern opposition to the war had reached its zenith. The North had teetered at an irrevocable tipping point. Fueled by fiery newspaper editorials, rank and file citizens increasingly demanded an end to the bloody conflict. Nowhere were such demands more vocal than in New York City. In fact, the city had been divided from the outset. Its wealthy merchants and bankers had been loath to lose their southern trading partners and customers. The city mayor, Fernando Wood, had called for New York City to secede from the United States.

Meanwhile the city's working class citizens viewed the war as benefitting only the rich. Adding insult to injury, the U.S. Congress had passed a conscription law, a military draft. All males between the ages of 20 and 45 were eligible to be called for military service. Yet, any person drafted could pay 300 dollars and be exempted.

Tempers flared, and on 13 July 1863 rioters took to the streets of New York. Their four-day rampage would become the largest civil insurrection in American history. The rioters burned government buildings, burned churches and an orphanage, and looted stores. City police were unable to quell the uprising, so Abraham Lincoln sent U.S. Army troops into the city. When the army finally drove insurrectionists out of the streets, 119 people had been killed. Over 300 more had been injured.

If Jackson had survived, would his reputation as an invincible warrior-general have further swayed public opinion? Would his presence on battlefields have forced northern politicians and the public to rise up and demand an end to the unpopular war? At the Battle of Gettysburg in July 1863, if Jackson had commanded Second Corps, would his presence have produced a different outcome? If the South had won there – which it almost did *without* Jackson – historian Tony Horwitz explains the inevitable outcome:

> The already demoralized Army of the Potomac would have fallen apart. Then, with the victorious Confederate Army on the loose in Pennsylvania and threatening northern cities, it would have been game-over for the Union.

Several historians have published scholarly "alternative history" novels that feature Jackson's survival and continued generalship. Two such works are *By Force of Arms* and *Stonewall Jackson at Gettysburg*. Also, *How Few Remain* portrays Gen. Jackson as the commander of an independent C.S. Army in the 1880s. Of course, all such works are presumptive fiction.

Jackson had died. His death encouraged the North to the extent that the irrevocable tipping point was never quite passed. The war continued. The terrible contest of arms would drag on for 23 more months, and hundreds-of-thousands more would fall on both sides. They, and their posterity that never would be born, are included among the tragic casualties of war.

> There is no instance of a country having benefitted from prolonged warfare.
> *[Sun Tzu, The Art of War, circa 500 BC]*

Jackson's Legacy: When Jackson had been two years old his father and sister had died. Five years later his mother also died, leaving her orphaned son to be raised by various other people. Yet, with no surviving brothers, sisters, or parents young Jackson had persevered by sheer force of will. In adulthood his fame had spread throughout Europe and America.

In the Third Millennium at VMI, first year cadets are required to salute at the Jackson Arch entrance to their barracks. Jackson's famed credo, "You can be whatever you resolve to be," is inscribed over the arch. On the VMI campus the Jackson Statue base is inscribed with Jackson's words spoken prior to the charge at Chancellorsville: "The Virginia Military Institute will be heard from today" (the two words, *Virginia Military*, were not actually spoken by Jackson, but they have been added in the inscription).

From Jackson's handwritten notes, his military lectures are still taught at VMI. Jackson had taught the timeless military leadership essentials: mobility; audacity; discipline; initiative; learning the enemy's intentions, strengths, and weaknesses; concealing your own plans and vulnerabilities; and coordination of infantry and artillery. In American and European armed forces, Jackson's battlefield tactics and mobility are still studied over 150 years after his death. In 1997 historian Douglas Gibboney wrote:

> Gen. Jackson was viewed as a magnificent eccentric who brought *[the South]* brilliant victories. His oddities only served to confirm his genius and set him apart from other men.

On three occasions (1936, 1970, and 1995) Jackson's likeness has been displayed on United States postage stamps. Three U.S. Navy ships including a nuclear ballistic missile submarine (*USS Stonewall Jackson*, SSBN-634) have been named for him. The lineage of his early command, the Stonewall Brigade, is alive and well as the U.S. Army 116th Infantry Brigade. Municipalities in eight states (Texas, Virginia, North Carolina, Alabama, Mississippi, Louisiana, Kentucky, and Oklahoma) are named "Stonewall" in his honor.

> His devotion to God, duty, and country remain treasured legacies of the American people.
> *[James Robertson, Stonewall Jackson, 1997]*

Gen. Jackson stands alone atop a high pedestal. Since 1970 the Virginia state senate has recognized his birthday each year. Dozens of statues have been erected to honor his memory. Numerous highways and schools bear his name. Along Interstate Highway 95 in Virginia, signs give motorists directions to the Stonewall Jackson Shrine (the small building in which he died).

However, Jackson still *lives forever* at one location in Georgia. East of Atlanta massive Stone Mountain rises to a height of 1,686 feet. It is an igneous quartz monzonite dome formed eons ago by upwelling magma from deep within the earth. The huge Confederate Memorial, the largest bas-relief sculpture in the world, is chiseled into the sheer north face of the mountain. It was begun in 1923 and completed 49 years later in 1972. The carved surface, recessed 42 feet into the stone, is the size of three football fields. The memorial depicts three men on horseback: Jefferson Davis, CSA President; Gen. Robert E. Lee, commander of the Army of Northern Virginia; and Gen. Thomas J. "Stonewall" Jackson, riding on Little Sorrell. The three men tower 400 feet above the ground. The memorial will last throughout the ages.

The Most Enduring Legacy: Jackson's most enduring legacy is not cast in bronze or sculpted in marble or granite. It lies in his manhood, not in his warrior ethos. In his adopted hometown the old lecture room at Lexington Presbyterian Church had been his classroom. There in 1856, deacon Jackson had organized and taught a Sunday School class for slaves and free Negroes. Exhibiting his staunch Calvinist faith, he led scores of Negro men, women, and children to Christian salvation. He also taught them to read. Teaching slaves to read violated Virginia law, but Jackson answered to a higher power. Like the fearless Joshua, he did not shirk his spiritual duty.

Lylburn Downing, a slave child, had been born on 3 May 1863 (the day after Jackson had been wounded). As he grew to adulthood after the war he often had listened as his parents talked of Jackson's tireless efforts to teach them to read. They also credited Jackson with their conversion to Christianity, and their gratitude and admiration inspired Downing. As an adult he completed theological studies at Lincoln University and became a Presbyterian minister.

In 1895, Rev. Downing set up a "mission" in nearby Roanoke, Virginia. Initially he had only seven members, but they proved to be the genesis for Fifth Avenue Presbyterian Church, a predominantly Negro house of worship. The church grew and prospered.

Rev. Downing never forgot his parents' devotion to Jackson. His congregation included some elderly former slaves who had attended Jackson's classes. Together they resolved to honor the man who had been so faithful to them. The idea of memorializing a Confederate general in a predominantly Negro church raised a few eyebrows, but the congregation remained steadfast.

> Not even fire could destroy bonds forged more than 150 years ago between a group of Virginia slaves and a Confederate general.
> *[Tonia Moxley, historian, 30 October 2011]*

Former slaves and slave descendants raised money to fund a stained glass window in memory of Gen. Jackson. Rev. Downing designed the window. It exhibits the Shenandoah Valley and Blue Ridge mountains. The window depicts soldiers camped by a river, stacked arms, and the words: "In Memory of Stonewall Jackson." Additional text paraphrases the general's deathbed statement.

Workmen installed the window behind the church pulpit, and the congregation dedicated it on 10 May 1906. The tenth day of May is the date of Jackson's death. The year 1906 is the 50th anniversary of the founding of Jackson's Sunday School class. For the dedication ceremony, congregation members were joined by other former slaves and slave descendants. Members of the local Confederate Veterans camp, all Caucasians, took part in the ceremony. Members of the press attended along with many prominent citizens in Virginia. The dedication prompted national newspaper publicity. Letters of praise came from as far away as England.

Rev. Vernie Bolden, a grandson of slaves, later served the church as pastor. He offered his perspective on the Jackson memorial: "It represents an ideal of what could be, and what should be, instead of the reality of what is." That sentiment, perhaps, is Thomas Jackson's most enduring legacy.

> Jackson fought for the constitutional rights of the South, and anyone who imagines he fought for slavery knows nothing of Jackson.
> *[William Chase, The Story of Stonewall Jackson, 1901]*

Photograph of Maj. Thomas Jackson, USA, circa 1848.

Photograph of Thomas Jackson as a professor at VMI circa 1855.

Photograph of LtGen. Thomas "Stonewall" Jackson, CSA, taken on 26 April 1863. Six days later on the night of his greatest military victory, he was accidentally shot by his own soldiers. He died of pneumonia eight days thereafter.

Stained glass window behind the pulpit at Fifth Avenue Presbyterian Church in Roanoke, Virginia. The window was designed by former slaves and slave descendants. It was placed in the church in 1906 in memory of Thomas "Stonewall" Jackson. In the 1850s, Jackson had organized and taught a Sunday School class for slaves, and he also taught them how to read.

The Thorny Issue of Slavery

> My paramount object in this struggle is to save the Union. It is not either to save or to destroy slavery. If I could save the Union without freeing any slave, I would do it; and if I could save *[the Union]* by freeing all the slaves, I would do it; and if I could save *[the Union]* by freeing some and leaving others alone, I would also do that. What I do about slavery and the Negro race, I do because I believe it helps to save the Union; and what I forbear, I forbear because I do not believe it would help to save the Union.
> *[Abraham Lincoln, U.S. President, in a letter to Horace Greeley, editor of the New York Tribune, 22 August 1862]*

Introduction: Generations born after the American Civil War hold a host of different opinions about the impact of chattel slavery on the war. Was the institution of slavery unique to the United States? Was slavery the primary issue that led to warfare? Or, were secession and the war prompted by other factors? Many historians and historical revisionists have painted different pictures of this issue. Therefore an emotionally detached examination of the economic, political, legal, and cultural status of slavery prior to secession is warranted.

Origins of Slavery: The word *slave* comes from the term *Slav*, for the first documented slaves were Caucasians from Slavic countries. Chattel slavery is a way of life in which people are held as *property* and forced to work without wages. Unless manumitted, a chattel slave remains a slave for life. As property, slaves could be bought and sold on the open market, and children born to slave parents would become slaves themselves.

Evidence of slavery became obvious after the development of agriculture around 12,000 BC to 8,000 BC. Slavery was the norm in all early civilizations: Assyria, India, Greece, the Roman Empire, the Islamic Caliphate, China, Egypt, Babylonia, Persia, etc. The right-of-conquest principle became established among nations and cultures. A country or military force that defeated another country or military force was entitled to spoils of war. The victor took captured but healthy soldiers and civilians as chattel property.

Viewed in the light of modern-day mores, slavery is history's greatest encroachment upon human liberty. Nonetheless, before and throughout the Middle Ages it was a socially accepted way of life worldwide. During the peak of their power the Romans held about 350,000 slaves in and around the city of Rome. Slaves made up roughly 35 percent of the population of Italy. These Eurasian slaves were Caucasians from conquered lands. One historian wrote:

> During the millennium from the emergence of the Roman Empire to its eventual decline, at least 100 million people were captured and sold as slaves throughout the Mediterranean and its hinterlands.

Beginning around 1300 AD, Barbary pirates on the northern rim of Africa entered the slave industry. They raided fishing villages on coasts of Portugal, Spain, France, and Italy and took men, women, and children as slaves. Muslim slave traders became notorious for their atrocities, cruelty, and inhumanity. They sold their slaves at Arab slave markets in the Middle East. Tartars made raids into southern Russia in a process they called "harvesting the steppe," and they captured thousands of slaves. They sold about 18,000 of them each year at the Crimean slave market in Kaffa.

Slavery had been a way of life for centuries in central Africa. Sub-Saharan jungles were heavily populated by tribal groups that vied for slaves and dominance. In 1441, tribal leaders in what would become present-day Mauritania expanded their business. From coastal depots they sent shiploads of slaves to Lagos, Portugal, and set up a slave market there. Their trade thrived. Within 50 years Negro slaves, sold by Negro traders, would constitute 12 percent of the population of the Portugese capital city, Lisbon.

> Unfortunately, these facts about slavery are not in the lessons taught in our *[American]* schools and colleges. Instead there is a gross misrepresentation and suggestion that slavery was somehow a uniquely American practice.
> *[Walter E. Williams, editorial, 19 July 2017]*

<u>European Slavery Comes to America</u>: In America the Spanish set up colonies in Hispaniola, Puerto Rico, Cuba, and throughout Latin America. They initially took local natives (aka, *Indians*) as slaves. Later both Spanish and English colonists brought Caucasian slaves from Europe to what eventually would become the United States. However, Caucasian slaves were expensive, and Indian slaves were prone to escape and melt back into the native population.

In the West Indies and in continental North America and South America, plantation owners wanted vast numbers of slaves to clear fields and labor on sugarcane farms and in mines. African slave traders filled this need. They induced Portugese and Spanish ship captains to sail directly to the slave-rich Gold Coast of western Africa. Soon Dutch, French, and English captains joined the rush to Africa to buy slaves for shipment to America. Local Negro trappers raided tribal regions and abducted men, women, and teenagers. They took them to the seacoast where they met with ship captains and negotiated a sale price.

The first slaves from Africa had arrived in Hispaniola in 1501, only nine years after Christopher Columbus had first landed there. Owners found Negro slaves to be more economical than Caucasian slaves and more reliable than Indian slaves. The African slave trade boomed. The *Trans-Atlantic Slave Trade Database* is considered the "gold standard" in its field. It shows that between 1525 (records for earlier years are not available) and 1866 about 12.5 million Negro slaves were shipped toward the Western Hemisphere. Of those, 10.7 million survived the voyage. Most landed in the West Indies, Latin America, and South America. Brazil alone received 4.86 million, close to half of the total. Only 388,000 Negro slaves, less than four percent of the total, landed in what later would become the United States. One historian explained:

> In the West Indies and South America in particular, but also in North America, slavery was the engine that drove the mercantile empires of Europe. It appeared, in the Eighteenth Century, as universal and immutable as human nature.

Vast sugarcane plantations in the West Indies fueled a booming slave trade. In Barbados in 1734 the slave population was estimated to be 86,546. Fifty-three years later the slave tally had soared up to

210,894. Nearby Jamaica was home to a quarter-million slaves. In adjacent Cuba an 1847 census enumerated 496,000 slaves.

Early Years of Slavery in the United States: The first 20 Negroes had landed in Virginia during 1619. They had come from Europe, and they were not slaves. They were indentured servants. Within 10 to 15 years most of them had served their contract term, had become freemen, and were farming on their own land.

A flood of slaves had followed the indentured servants. A Negro freeman, Anthony Johnson, had become the first man in Virginia to own Negro slaves. In 1725 the slave population in American colonies had been estimated to be 75,000. By the Revolutionary War the count had risen to a half-million. At that time slaves made up about 39 percent of the population of southern states. However, in northern states the number of slaves stood at only about two percent of the total population.

Southern delegates to the Constitutional Convention in 1787 wanted each slave to be counted as *one person*, because members of the House of Representatives and the Electoral College would be based upon state populations. Northern delegates opposed this plan. They wanted only "free persons" to be counted for apportionment purposes, but finally a compromise was reached. Each slave would be counted as three-fifths of a person.

Before development of Eli Whitney's cotton gin in 1794 most slaves in southern states worked on large tobacco plantations. Others worked as household servants, carpenters, blacksmiths, bakers, couriers, and stevedores. Yet, the newfangled cotton gin changed life in the South. Within a few years enormous cotton plantations created a market for additional thousands of slaves.

After the Louisiana Purchase in 1803 the slavery issue became both geographic and political, for 828,000 square miles of territory had fallen into United States hands. The matter became even more murky in 1848 when victory in the Mexican war brought another 700,000 square miles under United States control. For over a half-century the U.S. Congress and Supreme Court would grapple with slavery issues in legislation and rulings such as:

– Missouri Compromise (1820).

– Creation of the Bureau of Indian Affairs (1824).

– Wilmot Proviso (1846).
– 2nd Fugitive Slave Act (1850).
– 3rd Fugitive Slave Act (1850).
– Compromise of 1850 (1850).
– Kansas-Nebraska Act (1854).
– Dred Scott decision (1857).
– Crittenden-Johnson Resolution (1861).

The average subsistence farmer in the southern United States did not need slaves. Further, he could not afford to house, clothe, and feed them. Yet, some owners of medium sized farms had invested in one or two. Such slaves usually lived in the home of their owner and his family. The owner required them to work without wages, but he otherwise treated them as quasi-family members. Despite family relationships, physical abuse of slaves sometimes occurred. Such abuse almost always came not from owners, but instead from paid and irresponsible overseers.

> Regard an overseer as an indispensable agent whose first qualities should be honesty and fairness, united with forbearance and good temper.
> *[editor, Farmer's Register, 1873]*

Deed books reveal that slave prices in the mid-1800s would vary from around $300 to over $1,500 depending upon the age, sex, health, and skills of the slave. Such prices equate to present-day values from $4,000 to well over $10,000. No sane owner would intentionally allow abuse of such a valuable investment.

In the United States federal law had banned further importation of chattel slaves into the country after 1808. However, interstate use and sale of slaves remained legal. Slaves could be bought, sold, or given as gifts. For example, one deed names slaves to be conveyed:

> . . . *[specified real estate]* and also the following Negroes, viz: Josh, Charles, Rose, Jane, Van, Sandy Jr., Joseph, Isaac, Peter, George, Rena, Chany, Dianah, Jim, and Mack, with their future increase . . . .

Most churches counted slave owners, slaves, and Negro freemen among their members. Most slave owners were Caucasians, but there were exceptions. In addition to the previously mentioned Anthony

Johnson, William Ellison was a Negro merchant and a successful entrepreneur. Per the 1860 Census he owned 63 Negro slaves.

> Churches were the center of intellectual life in the South. That was where people congregated, where they learned about the world and their place in it, and where they received moral guidance.
> *[Gordon Rhea, in an address to the Charleston Library Society, 25 January 2011]*

Clergy in the South defended the institution of slavery based upon their interpretation of the infallibility of the Holy Bible. They touted the Bible as the universal guide for moral issues. Commonly cited from pulpits were scripture passages in the Book of Leviticus that upheld buying and selling of slaves. Also the Biblical admonition for slaves to obey their owners – written in the context of Biblical times and mores – often was quoted:

> Slaves, obey your earthly masters with respect and fear, and with sincerity of heart, just as you would obey Christ.
> *[Ephesians 6:5 NIV]*

By 1860 such pulpit validations of slavery had become strident. Of course, it must be acknowledged that the clergy had their own axe to grind. Their financial future was dependent upon maintaining the cultural status quo. Therefore southern churches denounced the North as sinful and decadent because northerners supposedly had rejected the Bible. According to the clergy's reasoning, because the North was sinful the South could purify itself only by secession.

During the past decade a novel, *Uncle Tom's Cabin*, had been written by a fiery abolitionist from Connecticut. Although it was a work of fiction, it was well written and by 1860 had become the best selling novel of the Nineteenth Century. The fictitious book was filled with heroes, villains, and a long-suffering Uncle Tom and his fellow slaves in the South. The novel painted all slave owners as fiends and further intensified the rift between North and South.

> Oh, I wish I was in the
> Land of cotton . . . .
> *[Daniel Decatur, Dixie, 1859]*

Slavery and the American Civil War: By 1860 the institution of slavery was widespread in the United States. The majority of slaves were found in southern states. The 1860 Census listed 3,500,000 slaves in states that would become the Confederacy, and 420,000 slaves in the rest of the United States. In the South most slaves worked in the cotton or tobacco industry on big plantations. Most citizens in the North had no *moral* objection to the institution of slavery. Actually, northern factory workers feared manumission of slaves. They reasoned that slaves, if freed, would be willing to work for lower wages and thus threaten northern job security.

Slavery, both in the North and in the South, was protected by the U.S. Constitution and federal laws. Article IV, Section 2, Clause 3 of the U.S. Constitution is known as the Fugitive Slave Clause, but it also applied to indentured servant contracts. It is quoted below:

> No person held to service or labour in one state, under the laws thereof, escaping into another *[state]* shall, in consequence of any law or regulation therein, be discharged from such service or labour, but shall be delivered upon claim of the party to whom such service or labour may be due.

For enforcement purposes the subsequent Fugitive Slave Act of 1793 had done four things: (1) created the mechanism for recovery of indentured persons and runaway slaves, (2) overruled any state law giving sanctuary to such runaway fugitives, (3) made it a federal crime to assist runaway fugitives, and (4) allowed slave-catchers to enter any state and any United States territory.

A decade prior to the American Civil War some abolitionist groups had tried to undermine the constitutional Fugitive Slave Clause and also the Fugitive Slave Act of 1793. Their efforts had failed, and Congress had passed the even more restrictive Fugitive Slave Act of 1850. It specifically required officials in all states to return fugitive runaways to their lawful owners.

The famous *Dred Scott* Supreme Court decision in 1857 again had affirmed the legality of slavery and denied citizenship to slaves. Also, early during the war on 25 July 1861, Congress would pass the Crittenden-Johnson Resolution. It would confirm that the war was being fought to preserve the Union, not to abolish slavery. In part the resolution text reads:

> Article 2: Congress shall have no power to abolish slavery in places under its exclusive jurisdictions and situate within the limits of States that permit the holding of slaves.
>
> Article 3: Congress shall have no power to abolish slavery within the District of Columbia.
>
> Article 4: Congress shall have no power to prohibit or hinder the transportation of slaves from one State to another.

When campaigning for the presidency in 1860, Abraham Lincoln did what successful politicians always do. He told potential voters what they wanted to hear. A majority of citizens in the North accepted slavery as a societal norm, and with this type of audience Lincoln would not address the issue. Conversely, when speaking to abolitionist groups he extolled the evils of slavery and called for its demise. Lincoln knew the public would not support abolition as justification for war. For example, in his inaugural address on 4 March 1861 he stated:

> I have no purpose, directly or indirectly, to interfere with the institution of slavery in the states where it exists. I believe I have no lawful right to do so, and I have no inclination to do so.

For Lincoln to have claimed otherwise during early years of the war would have courted disaster, so he explained that the purpose of the war was to preserve the Union. One of his greatest military leaders, Gen. Ulysses S. Grant, made his position clear in writing:

> If I thought this war was to abolish slavery, I would resign my commission and offer my sword to the other side.

As the war dragged along and draft riots raged in the streets of some northern cities, Lincoln desperately needed a unifying cause. Taking a tremendous risk, he issued the Emancipation Proclamation effective 1 January 1863. It named ten states in which former slaves would be "forever free." The proclamation was not a law passed by Congress, but instead an order based upon Lincoln's authority as commander-in-chief of the U.S. Armed Forces.

The proclamation did not apply to any northern state or to the "slave states" or "border states" that had not seceded. It made slaves in the Confederate States *freedmen* if the North won the war. The order lacked authority to make them United States citizens. Yet, with cooperation of the liberal northern press, Lincoln made eradication of slavery *in the Confederacy* a moral imperative. However, state laws in Kentucky, Missouri, Maryland, and Delaware still allowed slavery. These four "border states" had not seceded and had remained part of the Union. In these states, slavery would remain legal until the Thirteenth Amendment became effective in December 1865.

Ironically, although the Confederate government had allowed slavery to continue within its borders, it had banned importation of more African slaves from any country other than the United States.

> 1. The importation of Negroes of the African race, from any foreign country other than the slaveholding states or territories of the United States of America, is hereby forbidden . . . .
> *[Constitution of the Confederate States of America, Article I, Section 9, (1), 11 March 1861]*

Gradual Emancipation: Earlier during the war in August 1861, MajGen. John Fremont, a Union commander, had declared a state of martial law in Missouri. His subsequent military order had included emancipation of all slaves in the state. President Lincoln had been infuriated. He had rescinded the order and had removed Fremont from his command. Always the consummate politician, Lincoln explained his action in writing:

> I think there is a danger that the liberating of slaves . . . will alarm our southern Union friends and turn them against us *[and]* perhaps ruin our fair prospect for *[military and political success in]* Kentucky.

Lincoln feared that the "border states" that had not seceded, and in which slavery was widespread, would secede from the Union rather than allowing their slaves to be freed.

The next year in May 1862 another Union commander, MajGen. David Hunter, had inexplicably ordered freedom for all slaves in Florida, Georgia, and South Carolina. Once again Lincoln rescinded the military order. In addition to concern about possible border state secession, he worried that tens-of-thousands of newly freed slaves would create a massive refugee crisis. Slaves were accustomed to food and housing provided by their owners. If freed, where would they live? Who would feed them? Lincoln had decided that new colonies for former slaves, outside the United States and Confederate States, would be in the best interests of all parties.

In August 1862, Lincoln had met with Negro leaders in the North. He had suggested that with government assistance freed slaves could set up colonies in Latin America. He had promised that the United States would support and protect such colonies until they became self-sufficient. However, Negro leaders had rejected that concept and had explained that freed slaves would prefer to remain in Union controlled territory. Nonetheless, Lincoln had persisted in plans to establish colonies. He sent Negro colonists to preserves in Haiti and Latin America. However, history proved that none of these colonies would be able to sustain itself.

By January 1865 after his fabled march to the sea, Gen. William Sherman and his Union Army occupied Savannah, Georgia. Outside the city a vast refugee camp of about 12,000 former slaves lived in hastily erected tents and shanties. They relied upon the Union Army to feed them, a situation that could not continue indefinitely.

On 11 January the U.S. Secretary of War, Edward Stanton, arrived in Savannah and conferred with Gen. Sherman. The next day the two men summoned a group of 20 local Negro leaders, all of whom were ministers, mostly Methodist and Baptist. Eleven of them never had been slaves. They had been freemen from birth. Sherman asked the ministers: "What do your people want from me?" The group spokesman, Rev. Garrison Frazier, told Sherman that freed slaves wanted title to their own land. Most of the former slaves had been engaged in farming, and with their own land they supposedly could be self-sufficient.

In response, four days later Sherman issued Special Field Order No. 15. He ordered federal confiscation of a 30-mile-deep strip of coastal lowlands and coastal sea islands. The confiscated land would begin at Charleston, South Carolina, and extend southward along the eastern seacoast, across Georgia, all the way to the St. Johns River in Florida. This land would be stripped from its legal owners, both large plantation owners and small subsistence farmers alike. In part the order is quoted below:

> *[The confiscated land is]* reserved and set apart for the settlement of the Negroes now made free by the acts of war . . . . In the settlements hereafter to be established no white person whatever, unless military officers or soldiers detailed for duty, will be permitted to reside . . . . Each family will have a plot of not more than (40) forty acres of tillable ground.

"Forty Acres and a Mule": About 900 Negroes quickly moved to large Skidaway Island near the Georgia coast, got 40-acre tracts of land, and set up a community. Rev. Ulysses Houston became the self proclaimed "black governor." By the summer of 1865 a virtual flood of about 38,000 Negroes got 40-acre tracts and settled on 400,000 acres of "Sherman land" within the reserved strip. Soon the U.S. Army began lending mules to the new settlers to help them plow and till the soil. This practice led to the phrase, "Forty acres and a mule."

However, none of the settlements were destined to last. Later that year the new U.S. President, Andrew Johnson, rescinded Sherman's military order and restored the land to its lawful owners. Yet, even before Johnson's mandate, problems had arisen. Some 40-acre tract recipients were found to have been freemen all along. They never had been slaves. Other recipients had gotten tracts, but never had intended to live there and farm the land. They promptly sold the land to the highest bidder. The *promised land* concept withered and died. Only a few previously uninhabited sea islands remained in the hands of freed slaves. Further, one by one the new Negro colonies in the Caribbean and in Latin America were abandoned.

The only American-sponsored Negro colony that survived was Liberia, but its origin was unrelated to the war. Almost a half-century earlier in 1817 the American Colonization Society had been founded in Virginia. This philanthropist society had been formed to induce and assist Negro freemen resettlements in Africa. In January 1820 the first shipload of 88 freemen had set sail, and they set up a colony on the western African coast. Through 1835 six more freemen colonies had been established nearby. Three years later they had combined to create the Commonwealth of Liberia, with Monrovia as their capital city. Over passing years Liberia struggled, but it would survive with financial assistance and private industry investments from the United States and western Europe.

Demise of Slavery in the United States: After the war the Thirteenth Amendment to the U.S. Constitution, ratified in December 1865, outlawed slavery and other forms of involuntary servitude except as punishment for a crime. The amendment reads in part:

> Section 1: Neither slavery nor involuntary servitude, except as punishment for *[a]* crime whereof the party shall have been convicted, shall exist within the United States, or any place subject to their jurisdiction.

Despite legal mandates in the United States and elsewhere, chattel slavery did not die overnight. As late as 1900 almost two million slaves remained in the Sokoto Caliphate in Africa. In 1908 female slaves still were bought and sold at markets in the Ottoman Empire. Twenty years later Ethiopia was home to around 900,000 slaves. In 1964 Saudi Arabia finally made slavery illegal. The last country in the world to outlaw slavery was Mauritania, which criminalized the trade in 2007. Nevertheless, despite its legal prohibition, slavery would remain common in parts of Asia and in regions of Cameroon, Gabon, Nigeria, Togo, and Guinea in central Africa.

In the present day some people assume that all Negroes in the United States are descendants of slave families. Of course, such is not the case. Many Negroes and their ancestors were freemen who lived in the mainstream of society. For example Henry Brown, a Negro, was born a freeman in Camden, South Carolina. Neither he nor his ancestors ever were slaves. At age 18 he had joined the U.S. Army and had served during the Mexican War. Thereafter he became a brick mason and community leader. After his state seceded from the United States in 1860, Brown had joined the Darlington Grays, 8th South Carolina Infantry Regiment, C.S. Army.

After the bloody war Brown would join the Darlington Guards, a Confederate veterans organization. In later years he would receive a Confederate Veterans pension from his state. After he died on 2 November 1907 the *Darlington Press* would describe him:

> In life he was faithful to every trust, and his word was his bond.

Slave Descendants in the United States: Henry Brown, who never had been a slave, was in the minority. The majority of Negroes in southern states had been slaves or descendants of slaves. Although slaves technically had been manumitted after the war, it would take over 100 years before some slave descendants joined the mainstream of American society.

Many civic organizations such as the National Association for the Advancement of Colored People (NAACP) would become ardent advocates for racial equality. Despite such efforts some uneducated descendants of former slaves would remain beguiled by well-heeled purveyors of a cruel *victimization industry*. Thus captivated and controlled, they would cling to many generations of mediocrity and financial dependence upon government.

Conversely, in the United States middle-class and upper-class descendants of former slaves realize that their ancestors' chattel status got them out of Africa. Although slavery is totally unacceptable by modern standards, it nonetheless had gotten them and their posterity out of a region where famine, poverty, violence, a hand-to-mouth existence, and slavery still exist.

By the Twenty-First Century educated slave descendants in the United States realize that they enjoy the fruits of a stable economic system, government safety nets, workplace equality, free education for their children, professional health care, and personal liberties unparalleled anywhere else in the world. Consequently they have assigned the chattel status of their distant ancestors into a closed chapter in their history books.

> The rose-colored myth of the Civil War is that blue-clad Union soldiers and their brave leader, Abraham Lincoln, were fighting to free the slaves. They weren't . . . . They were fighting to hold the Union together.
>
> The Emancipation Proclamation *[freed]* slaves in the rebel states, but not in the border states, which Lincoln needed to remain loyal to the Union.
>
> *[Sarah Pruitt, Myths About Slavery, 2016]*

Etching depicting Muslim slave traders negotiating the sale of Caucasian slaves at an unidentified seaport in 1637.

Oil painting, The Slave Market by Gustave Boulanger, depicting a Roman slave auction of children and young adults.

Etching depicting slave traders negotiating sale prices at Goree, Senegal, during the 17th Century.

Three slave children photographed in New Orleans in 1863.

Photo of a watercolor painting that depicts slave-catchers in Brazil. Of the estimated 10.7 million Negro slaves who reached the Western Hemisphere, 4.86 million of them landed in Brazil.

"Uncle Marian," an elderly slave in North Carolina.

Four generations of a slave family at Beaufort, South Carolina, in 1862.

Nautical Warfare

A man-of-war is the best ambassador.
[Oliver Cromwell (1599-1658), Lord Protector of Ireland]

Introduction: This project concentrates upon generational human losses and core issues during the war. Military events are not the primary focus, for many thousands of books and documents already have addressed such issues in great depth.

However, relatively little has been written about naval forces of the Confederate States. Most laymen have no inkling of the enormity of military naval operations during the war. Therefore this chapter highlights the C.S. Navy, civil privateers, and blockade runners.

Evolution of Naval Warfare: During antiquity and the Middle Ages the Romans, Greeks, Egyptians, and Persians built fleets of military galleys. These long narrow ships were powered by ranks of oarsmen. For long sea voyages the oarsmen sometimes were supplemented by a single sail. These vessels were designed to sink enemy ships by ramming them. Alternatively they could close and bind themselves to an enemy ship. This would allow their boarding parties to attack enemy crews in hand-to-hand combat.

The first naval *artillery* came into use during the Third Century BC. It consisted of shipboard catapults that could hurl missiles and incendiaries at enemy ships and their crews. Catapults soon were supplemented by huge deck-mounted crossbows that could fire giant arrows (aka, *bolts*) at opposing vessels.

Ramming tactics fell out of favor during the subsequent age of sail. Large shipboard naval cannons became the cutting edge of military technology around the Fourteenth Century. Heavily armed men-of-war emerged over the next 200 years. Such warships featured rows of large caliber cannons on "gun decks" along each side of the ship. The preferred tactic when attacking an enemy vessel required a heading of 90 degrees from the target. Then a "broadside" could be fired from all guns on the side of the ship facing the adversary.

The Nineteenth Century witnessed revolutionary changes in naval warfare. Around 1825, steam-powered engines and sidewheels or propellers (aka, *screws*) began augmenting sail power. Ships could cruise the seas using steam power alone, or sail power alone. They also could use both simultaneously. Beginning in the 1850s almost all new warships used steam engines as one, if not the only, means of propulsion. The advent of explosive shells led to use of protective iron plating to armor the deck and sides of warships. The first two fully ironclad ships, the British *Warrior* and French *Gloire*, signaled the coming obsolescence of wooden men-of-war.

Old military naval terms slowly died. New naval behemoths became known as battleships or ships-of-the-line. New rotating turrets allowed larger guns which could be aimed in any direction, regardless of the ship's heading. Older and smaller warships could not survive if they traded fire with new and larger ships-of-the-line. The smaller wooden vessels would continue in use as merchant ship escorts, scouts, river and harbor patrol, and coastal blockade duty.

Confederate Privateers: After hostilities began at Fort Sumter the South wasted little time in efforts to attack United States merchant shipping. On 17 April 1861, president Jefferson Davis issued a written proclamation to prospective privateers:

> I, Jefferson Davis . . . invite all those who may desire, by service in private armed vessels on the high seas, to make application for commissions or letters of marque and reprisal, to be issued under the seal of these Confederate States . . . .

Since the early days of sail, privateering had been an honorable profession and a way for nations to supplement their naval strength. A nation could issue a "letter of marque and reprisal" to a person, usually a ship owner or ship captain. This commission authorized the holder to conduct maritime hostilities against foreign vessels, usually unarmed merchant shipping. Captured ships and their cargo would be brought before admiralty tribunals for condemnation and sale. Proceeds of the sale would be shared by the named privateer owner, the captain and crew, and the nation that issued the commission.

Privateering had become common among European powers by the Sixteenth Century, and it was highly profitable. For example, a privateer who captured a Spanish galley laden with South American gold would become wealthy almost beyond measure. Privateers were similar to pirates, but there were two crucial differences between them. A pirate commander and his crew did not have to share their captured booty with anyone. However, they had no legal protection, and if they were captured the usual penalty was death by hanging.

The typical privateer ship was fast, lightly armed, but heavily crewed. The vessel owners were investors who stood to make handsome profits. Privateer ships operated in much the same manner as military commerce raiding ships of a legitimate navy. The main difference, of course, is that regular navy commerce raider crews did not reap financial proceeds when an enemy ship was captured.

Privateers enjoyed their greatest success during the first two years of the war before the U.S. Navy had enough warships to partially protect Union merchant shipping. However, often the weather and simple bad luck took a lethal toll on privateer sailors-of-fortune.

The privateer *SS Jefferson Davis* was a 187-ton brig. In over a month at sea she captured nine merchant ships. However, she tried to sail into the harbor at St. Augustine, Florida, during a storm and ran aground. The ship was a total loss. The privateer *SS Savannah* left port on 2 June 1861. On her second day at sea she captured the cargo brig *SS Joseph*. That afternoon she spotted another ship and chased it. Unfortunately, too late she discovered that her intended prize was not a merchant ship. Instead it was a warship, *USS Perry*. After a brief cannonade the U.S. Navy sank the privateer and captured Capt. Harrison Baker and his crew. Another bold privateer, *SS Petrel*, left Charleston on 28 July. That same day she blundered into the frigate *USS Lawrence*. Her captain and crew became captives.

Despite legal precedents to the contrary, the Union announced that it would treat captured privateer crews as pirates. In fact, one captured crew was tried for piracy in New York, but the jury failed to convict the mercenary sailors. In the end no captured privateer crewmen ever were hanged, for the Confederacy had promised to retaliate in kind against Union POWs.

The Fledgling C.S. Navy: Before the outbreak of hostilities in April 1861 the Confederate government had known the U.S. Navy likely would blockade southern ports and mouths of major rivers. To defeat or evade this blockade the C.S. Congress established the C.S. Navy on 21 February 1861. This new navy had three objectives:

– Protect southern harbors and coastlines from invasion.
– Attack United States merchant shipping worldwide.
– Evade or attack the blockade to ensure that shipping of arms, essential military supplies, commercial goods, and mail to and from Europe could continue.

The new Confederate Navy had 21 steam-powered vessels. None were warships. The Confederacy would have to build military ships or purchase them in Europe. The Confederate president, Jefferson Davis, appointed Stephen Mallory as Secretary of the Navy. Mallory, a respected admiralty attorney and former chairman of the U.S. Senate Naval Affairs Committee, faced a daunting and unenviable task. The South lacked enough shipyards, raw materials, industry, and expertise necessary to build modern warships quickly.

In contrast the U.S. Navy had 62 warships of various sizes in its fleet. Yet, it would have to construct many more frigates before a blockade could be moderately effective along 3,500 miles of Atlantic coastline. There were almost 200 usable harbors and navigable river mouths – down the east coast, down and around Florida, westward along the Gulf coast as far as Mexico, and up the Mississippi River. Further, Texas shared a tremendous open land border with Mexico, so a blockade of any type there would be impossible.

Mallory realized the Confederate States could never match the industrialized North in a ship-building race, so he opted for new technology, ironclad ships. He wrote:

> I regard the possession of an iron-armored ship as a matter of first necessity. Such a vessel at this time could traverse the entire coast of the United States, prevent all blockades, and encounter, with a fair chance of success, their entire navy. . . . Inequality of numbers may be compensated by invulnerability. . . . Naval success dictates the wisdom and expediency of fighting with iron against wood.

At the start of hostilities the South faced overwhelming shortages of military necessities. The new C.S. Navy had a limited amount of coal on hand, and the only domestic sources were in remote regions of Alabama and North Carolina. The C.S. Army faced severe shortages of military stores. All tools of war were in short supply.

Fuses and percussion caps were scarce, and there was no factory in the South that could produce them. The outlook appeared dire. In April 1861, Gen. John Grayson wrote to president Davis:

> As sure as the sun rises, unless cannon, powder, etc. be sent to Florida in the next thirty days, she will fall into the hands of the North. Nothing human can prevent it.

Quest for Ironclad Supremacy: The largest shipyards in the South were at Norfolk, Memphis, and New Orleans. Yet, ships built there had been designed for river commerce, not for battling U.S. Navy warships. Perhaps an ironclad ship could be constructed in one of these shipyards, but that would take time. The few scattered iron mills in the South had gotten their iron ore from Pennsylvania in the past, but that source was no longer available. Perhaps a suitable ironclad warship could be purchased abroad.

Mallory dispatched James Bulloch to Europe. He was a former U.S. Navy officer who had resigned and cast his lot with the South. The new French ironclad *Gloire* had cost twice as much as a similar wooden frigate, but it immediately caught Bulloch's eye. It seemed that the warship might be for sale, so he wrote to Mallory:

> This seems like a large price to pay for a six-gun ship. The finest wooden frigates afloat, carrying 40 guns of the heaviest caliber, cost but half of this amount. But no comparison of their relative values *[can]* be made inasmuch as the most formidable wooden frigate would be powerless in a contest with such a ship.

The C.S. Congress authorized $2,000,000 for purchase of *Gloire*, but France refused to sell her. Since no other ironclad warship was for sale, the C.S. Navy decided to have new ironclads built by English shipyards. Ownership of these new vessels was to be cloaked in ambiguity. Two new ironclad rams plus an ironclad frigate were to be the nucleus of a C.S. Navy ironclad fleet. Such a fleet could decimate the U.S. Navy blockade. Bulloch also ordered six steam-powered wooden ships that would be built at Liverpool, England.

The two small ironclad rams were built, but their seaworthiness left much to be desired. They managed to complete the trip across the Atlantic, but thereafter they were used only for harbor defense. In this role they served admirably.

Confederate efforts to finance additional ironclads in Europe were frustrated by financial problems. British shipyards initially were eager to extend credit to the Confederacy. However, eventually a shortage of funds doomed Confederate efforts to build more ironclad ships overseas. Nonetheless the South was able to purchase several new wooden steam-powered commerce raiders. These speedy ships would wreak havoc on Union merchant shipping during the last two years of the war.

The South was forced to use its domestic resources in its quest to obtain more warships. By August 1862 the C.S. Navy had set up two engine, boiler, and machine shops plus five ordnance shops at its shipyards. The navy also built a rope-walk, a factory that made nautical ropes and iron cables. Thus equipped, southern shipyards began building wooden warships and floating batteries.

Domestic Warship Production: Shipyards throughout the South began devoting all resources toward warship production. The task proved difficult. Steam-engine manufacturing capacity was less than ten percent that of the North, but some new ships were outfitted using this limited capability. In addition the South purchased existing merchant steamers and stripped them of their engines and machinery for use in new warships.

Many soon-to-be ironclad warships, but lacking the metal armor plates, eventually would be built in the South. Builders sent armor specifications for each such ship to England. English shipyards then manufactured the required armor plates and shipped them to the Confederacy via blockade runners. Upon receipt, workers in southern shipyards installed the iron armor.

Of necessity southern shipyards were located along the seacoast or along major rivers. As the war dragged along most of these shipyards would fall, sooner or later, to the advancing U.S. Army and U.S. Navy. For example, when the U.S. Navy captured New Orleans and pushed upriver to capture Memphis, huge shipyards in these two cities were lost. Time after time at these and other shipyards in the South, Confederate warships under construction would have to be abandoned, or burned and scuttled to prevent their capture.

In the South the greatest source of warships proved to be ships that had been captured when the war began. When the South had occupied United States harbors and shipyards in 1861, seaworthy military vessels had been seized by the Confederacy. These ships became part of the new C.S. Navy. Civil vessels under construction also were seized and adapted for military use.

Blockade Runners: Blockade runners were seagoing steam-powered cargo ships designed to elude the U.S. Navy blockade of southern ports and rivers. These vessels were designed for speed, and most had been built in England or France. Some were screw-driven, while others used sidewheels. Most were privately owned, although nine were C.S. Navy vessels. All of them ferried cargo to the Confederacy from neutral ports in Bermuda or the West Indies. On return trips they carried cotton, tobacco, and other trade goods.

In a typical operation, conventional and slower merchant ships would carry cargo from Europe to Caribbean ports. There the cargo would be offloaded and reloaded into swift blockade runners. These ships could carry a tremendous cargo load. For example, when the blockade runner *Fingal* (later converted to a big casemate ironclad warship, *CSS Atlanta*) made a run into the port at Savannah she delivered 400 barrels of gunpowder, 10,000 Enfield rifles, a million cartridges, and two-million percussion caps.

James Bulloch, the Confederate agent in Liverpool, arranged both ship construction and arms shipments. Through the English firm of Fraser, Trenholm, and Company he coordinated necessary financial transactions. London Armoury Company was the largest supplier of firearms of all types. By the end of 1862 it had shipped 70,000 rifles to the Confederacy. Most such munitions and armament from Europe passed through the port at Liverpool.

A West Point graduate, Maj. Josiah Gorgas, was a Confederate agent, and before the war he had worked in the U.S. Ordnance Bureau. Because of that experience he was picked to coordinate European arms shipments to the proper Caribbean port. Once the cargo arrived in the West Indies, John Maffitt, a Confederate naval officer and former privateer, would set up blockade runner shipment to a Confederate port.

When docked in a southern port, the average turn-around time for a blockade runner was about four weeks. First, the ship would have to be unloaded, a process that would take several days. Then she would have to re-coal. After that she would wait for her contracted outbound cargo to arrive. That could take a day or two, or it might take a few weeks. Contracts and manifests then could be prepared and signed, and loading of outbound cargo could begin. Then the captain would wait for a dark night with no moon before starting his nocturnal dash past blockading Union warships intent on sinking or capturing his ship.

Blockade running evolved into an art. The ships were steam-powered, long and narrow, with shallow drafts that allowed them to slice efficiently through the water. Most were painted dark gray so that they would blend in with the sea and sky at night. When steaming several hundred miles across the sea their boilers would be fired with normal coal, which produced a dense black smoke. When approaching the blockade they would switch to anthracite coal. It was smokeless and made the ship very difficult to see on the horizon. When speeding past the blockade the crew often would fire their boilers with cotton that had been soaked in turpentine. This resulted in more intense boiler heat and faster speed.

Successful blockade runners would approach a blockaded port only at night. Not every night would be suitable. The crew would wait for a cloudy or rainy night with no moon. When the crew extinguished all external lights on such a night, their ship would be almost invisible. In the dark with a full head of steam the stealthy blockade runner would zip past anchored U.S. Navy warships.

Blockading warships resting at anchor several miles from the harbor entrance were ill equipped to stop speeding blockade runners. Even if the intruding vessel was sighted, it would take time for the warships to set sail, or fire their boilers, and maneuver to bring their guns to bear. By that time the blacked-out blockade runner usually would have vanished into the dark night.

To supplement harbor and river blockade efforts, Union warships set up roving patrols in international waters near blockade runner ports in the West Indies. Such efforts met with little success. Blockade runners on the open ocean, unconfined by narrow harbor entrances, could slip past Union patrols at night. If they chanced to encounter a patrolling warship far out at sea in the daytime, they simply used their speed to outrun it.

During the first year of the war New Orleans was the largest cotton port in the world. Over 300 blockade runners used the river city as their home port. However, the U.S. Navy captured New Orleans in April 1862. The port at Mobile, Alabama, then became the most popular Gulf port for vessels running the blockade. Two years later in August 1864, Adm. David Farragut of the U.S. Navy made his famous "damn the torpedoes" (later called, *mines*) charge past Confederate forts and into massive Mobile Bay. Thereafter for the duration of the war, Galveston, Texas, became the preferred Gulf port for blockade runners.

Blockade runners leaving ports in Bermuda and the Bahamas usually made their runs to Charleston, Wilmington, or Savannah on the eastern Atlantic coast. The port at Wilmington became the largest Confederate coastal depot for military supplies during the war. A massive fortress complex, Fort Fisher, protected the Cape Fear entrance to the harbor. For over three years the U.S. Navy and Army tried unsuccessfully to capture the fort and stem the flow of arms to Wilmington. Finally in December 1864 an enormous U.S. Navy fleet tried to block the harbor entrance and capture Fort Fisher. The operation involved 123 warships and transports, the largest such attack force of the war.

> Blockade running seems almost as brisk as ever. The new class of blockade runners are very fast and sometimes *[they]* come in and play around our vessels. They are built entirely for speed.
> *[Adm. David Porter, USN, North Atlantic Blockading Squadron near Fort Fisher, North Carolina, in a letter to the War Department, December 1864]*

The attack took a month, but on 15 January 1865 the combined army and navy assault succeeded. Fort Fisher fell to Union forces. Ironically, during the final assault a blockade runner, *SS Wild Rover*, sped through the attacking Union fleet in broad daylight. She then steamed upriver to safety in the harbor at Wilmington. Three more blockade runners followed right behind her.

No one now knows how many vessels were involved in running the blockade during the war. The total is certainly in the thousands. The U.S. Navy estimates that it captured 1,100 such ships. Per the federal navy tally another 356 blockade runners were either sunk by military action or forced to run aground. How many successful runs were made? There is no official tally. The vessels were privately owned, and almost all individual ship records are now lost. However, records of *SS Syren* survived. She made 33 successful round-trips through the blockade during the war.

C.S. Marine Corps: Following naval tradition the C.S. Congress established the C.S. Marine Corps on 16 March 1861. Patterned after the U.S. Marine Corps, the new corps of Confederate Marines would be an infantry arm of the C.S. Navy. The Confederacy set up Marine Corps headquarters and training facilities at Camp Beall on Drewry's Bluff, Virginia, and also at the Gosport Shipyard.

Recruiting Confederate soldiers-of-the-sea posed no problem. Many men in the United States armed forces elected to "go south" with their home state, and the U.S. Marine Corps was no exception. Sixteen of its officers and about 200 enlisted men resigned and headed to Virginia to join the new C.S. Marine Corps. The transition could be difficult. Robert Tansill, a Virginian, was serving aboard the *USS Congress* as a captain in the U.S. Marine Corps. While aboard ship he resigned his commission, openly stating that he intended to join the C.S. Marines. The Union arrested him and jailed him until January 1862, when he was released in a prisoner exchange. Upon release he "went south" and was appointed to the rank of captain in the C.S. Marine Corps.

Congress had approved a force of 1,026 Marines, but manpower totals never exceeded about 539 men during the war. Yet, the small force ashore trained for special operations. At night a platoon rowed out to the anchored *USS Underwriter*, climbed aboard, and captured the ship. That scenario played out again when another contingent captured *USS Water Witch*. A nighttime raid to free prisoners of war at Point Lookout Prison Camp failed only after the Union Army learned of the plan in advance.

Under their new colonel-commandant, Lloyd J. Beall, other C.S. Marines began their primary duty, service aboard C.S. Navy ships-of-the-line and commerce raiders. The most successful was the *CSS Alabama*, for she sank or captured over $11,000,000 worth of Union merchant shipping. Marines also served aboard many river gunboats.

Ashore they offered security for navy bases at Norfolk, Wilmington, Charleston, Savannah, Mobile, New Orleans, and Galveston. During the last two years of the war they served as infantry and helped to defend coastal installations, most notably Fort Fisher in North Carolina and Fort Blakely in Alabama.

In the desperate early months of 1865 the C.S. Marines ashore helped to defend their capital city, Richmond. After the war ended in April the Confederate States Marine Corps would exist only in old documents and history books.

Unique Ships of the C.S. Navy: Stephen Mallory, the Confederate Secretary of the Navy, knew that most financial assets of the South would be used to support the C.S. Army. He would have a limited budget, so he developed a strategy to accomplish as much as he could with the resources at hand. First, he would purchase or build ironclad men-of-war for coastal and harbor defense and for attacks against the Union blockade. Next, he would buy or build cruisers suited for worldwide commerce raiding. Five unique Confederate warships are detailed below:

CSS Stonewall (aka, *Sphynx, Stoerkodder, Kotetsu, Azuma*): In warfare as in commerce, timing is crucial. There are many sad tales of too-late arrivals and lost opportunities.

Two years into the war the Confederacy still experienced great difficulty in buying or building new ironclad warships overseas. In June 1863, John Slidell, a Confederate emissary in France, obtained a private audience with Napoleon III, the emperor. French neutrality laws prevented building a warship for foreign belligerents such as the Confederacy. Yet, Slidell and the emperor reached an agreement. A new ship would be built, but identity of its purchaser would be kept secret. Also, no guns would be installed in France. The vessel would be named *Sphynx* to bolster false rumors that she was destined for the Egyptian Navy.

The warship keel was laid in Bordeaux Shipyard within a month. The ship's guns would be manufactured separately in England. The ironclad ram would be a nautical behemoth designed to sweep aside any opposing warship afloat. She was intended to wreak destruction upon the Union blockade of southern ports, harbors, and rivers. According to Union spies the dreadnaught would carry a crew of 135 and would be invincible in battle.

The new ship-of-the-line would differ from conventional ironclads. The heaviest gun would be a massive 300-pounder, plus smaller rifled cannons. Other ironclads of the era were converted wooden vessels, with armor later bolted over the wood for protection. However, the new Confederate warship was an ironclad monster from the beginning. She displaced 1,358 tons, was 194 feet in length, carried 95 tons of coal, and featured a double reciprocating steam engine. A heavy five-inch-thick layer of iron served as her main armor belt and covered her sides and turret. The iron ram on her bow mimicked a gigantic lance.

The U.S. Government was not fooled. It knew the new warship would attack the blockade. Using its considerable influence, the United States blocked the not-so-secret sale to the Confederacy. Not to be thwarted, the French sold the completed ship to Denmark. Renamed *Stoerkodder* and manned by a Danish crew, the ship embarked on a shakedown cruise on 21 June 1864. In England her guns and munitions were loaded aboard and mounted. Of course, the sale to Denmark was a ruse. When the ship arrived in Copenhagen on 6 January 1865 she was immediately manned by a waiting Confederate captain and crew.

Commanded by Capt. Thomas Jefferson Page, the ship left Denmark and headed for blue water. While at sea the captain re-commissioned the new C.S. Navy warship with its originally intended name, *CSS Stonewall*. She steamed to Spain for provisions and was shadowed en route by two U.S. Navy warships, *USS Niagara* and *USS Sacramento*. While *Stonewall* was in port the frigates waited outside the harbor entrance. One historian explained:

> On March 24, Captain Page steamed Stonewall out to sea, challenging the U.S. Navy vessels. They turned and fled, fearful of engaging the powerful new Confederate ironclad beast.

The two Union warships left without firing a single shot. *Stonewall* steamed westward and crossed the Atlantic. Capt. Page intended to attack the Union fleet at Port Royal, the supply port supporting Gen. William Sherman's attack into South Carolina. First he made a quick stop at Nassau, then stopped in Cuba to re-coal and take on water and more provisions. While in Havana he learned that the war already had ended.

CSS Stonewall never fired a shot in anger. Capt. Page surrendered his ship to the Spanish Captain General. Cuba then relinquished the vessel to the U.S. Navy. During the demilitarization that followed the war the U.S. Navy had no foreseeable use for the hi-tech *Stonewall*. In 1869 the U.S. Government sold her to the Imperial Japanese Navy, which was modernizing its battle fleet. At that time Japan was engaged in the Boshin War, which pitted Imperial forces against shogunate insurgents.

The Japanese renamed the ship *Kotetsu* (meaning, *iron ship*). She became the first ironclad vessel in the Japanese Navy. With a new Japanese Navy crew the ship steamed to Hokkaido to help battle Tokugawa forces trying to form an independent republic there. In later battles at Miyako Bay and Hakodate Bay where she led the Japanese Navy line of battle, *Kotetsu* proved to be truly invincible. In 1874 the ironclad monster was renamed *Azuma* (meaning, *East*) and fought as part of the Japanese fleet during the Taiwan Expedition. She ran aground during a typhoon but was fully repaired at the Yokosuka Naval Arsenal. Thereafter she stood guard duty in various Japanese ports.

Passing years took a technological toll. Newer warships featured exponential advances in gunnery, communications, hull design, propulsion, and defensive armament. Gradually the erstwhile Confederate monster became outdated. On 28 January 1888 she was decommissioned. Shortly thereafter Japan sold her to private investors, and she was scrapped.

CSS Shenandoah (aka, *Sea King*): *CSS Shenandoah* holds a unique place in Confederate naval history. She began life as a civil passenger and cargo ship, became a highly successful Confederate commerce raider, circumnavigated the globe, and fired the last shot of the war (across the bow of a whaler in the Bering Sea). She never dropped anchor in an American port and did not surrender until seven months after the war ended. Her story bears retelling.

The new civil vessel *Sea King* sailed out of the shipyard at Glasgow, Scotland, in August 1863. Designed for the East Asia tea trade, she was 230 feet in length. She was an iron-framed wooden triple-masted and fully rigged sailing ship, but she had a steam engine for auxiliary power. The 1,018-ton vessel made several cargo trips to the Far East, and she carried troops to New Zealand during the Maori War.

The next year in October 1864 the C.S. Navy bought *Sea King* under terms of absolute secrecy. Supposedly still a merchant ship, *Sea King* sailed from London. Ostensibly she was bound for India, but she rendezvoused with *SS Laurel* at sea. *Laurel* carried the nucleus of a C.S. Navy crew, naval cannons, powder, other military stores, and provisions. The new Confederate commander, Lt. James Waddell, converted the ship into a C.S. Navy man-of-war, the *CSS Shenandoah.* He lowered the Union Jack and raised the Stainless Banner of the Confederacy.

Lt. Waddell's mission called for him to "seek out and utterly destroy *[Union]* commerce in areas yet undisturbed." Waddell pursued his mission with vigor. He sailed south, rounded the Cape of Good Hope, and entered the Indian Ocean. When he encountered six Union merchant ships he took the crews captive. He burned five of the ships. The sixth ship he used to send captives to Bahia, Brazil, where they would be released. In January 1865, Waddell dropped anchor at Melbourne, Australia, where he replenished his stores and enlisted 40 more crewmen.

Leaving Australia, *Shenandoah* headed toward waters of the North Pacific. En route she captured four Union whaling ships near the Caroline Islands. Heading farther north, Lt. Waddell passed the Aleutian Islands and entered the frigid Bering Sea. There he captured eleven more Union whaling vessels. On 27 June he captured another whaler, the *Susan & Abigail.* Her captain showed Lt. Waddell a San Francisco newspaper that reported evacuation of Richmond two months earlier. However, the article included president Davis' order that "war will be carried on with renewed vigor."

Thus informed, Waddell continued to seize and destroy Union whalers in the rich Arctic hunting ground. He seized ten more ships and then headed southward toward San Francisco. He knew the city had almost no defenses, so he intended to attack and destroy its port facilities. Nearing the California coast on 3 August, he hailed the English barque *Barracouta.* From her captain, Waddell learned that the Confederacy had capitulated. Gens. Robert E. Lee, Joseph E. Johnston, and Kirby Smith had surrendered, and Jefferson Davis had been captured. The war was over. One historian explained the quandary facing Lt. Waddell and his crew:

> The crew knew that returning to a United States port would mean facing a Union court. News of president Lincoln's assassination served to further diminish any expectation for leniency. They could be tried and hanged as pirates. Commerce raiders were not included in the amnesty that Confederate Army soldiers were given.

Waddell lowered the Confederate flag. He dismounted his cannons and stored them out of sight below decks. His crew repainted the ship so that it looked like a civil merchant vessel. Waddell then sailed south past Mexico and South America, rounded Cape Horn, crossed the Atlantic, and headed for England, a three month voyage. He dropped anchor at the mouth of the Mersey River near Liverpool. There he once again raised the Stainless Banner.

At high tide with sails furled, *CSS Shenandoah* steamed up Victoria Channel and anchored near the British man-of-war *HMS Donegal.* There on 6 November 1865, Lt. Waddell surrendered his ship and crew to the Royal Navy captain and lowered the Confederate flag for the final time. Thereafter Lt. Waddell walked to Liverpool Town Hall and presented the mayor with a letter, giving *CSS Shenandoah* to the British government.

For the crew the saga had not ended. Their fate remained uncertain. However, after a formal Crown investigation the British government decreed that Waddell and his crew had not violated the laws of war or the laws outlawing piracy. They were to be set free. The *Liverpool Mercury* newspaper reported on 9 November:

> The crew were mustered on the quarterdeck by the officers of the ship, the roll book was brought out, and the names of the men *[were]* called out. . . . By order of the Government they were all paroled.

H.L. Hunley (civil submarine): When war began submersible vessels had seen limited civil use for several years. Most had been used for underwater salvage duties. After hostilities broke out during 1861 the U.S. Navy began construction of an experimental military submarine, *USS Alligator*.

In Louisiana a maritime enthusiast, Horace L. Hunley, would finance and build four civil submarines. All of them were intended to attack warships of the Union blockade. In New Orleans he built *Pioneer* and tested it on the Mississippi River and Lake Pontchartrain. Unfortunately, Union armed forces steadily advanced toward the river city, so Hunley scuttled *Pioneer*. He moved to relative safety in Mobile, Alabama, and began work on a clone, *Pioneer II*. Profiting from his experience to date, he abandoned the project and began work on a more advanced submarine, *American Diver*. In early 1863 he tested the new vessel, but it proved to be too slow for military use. Further, during an evaluation cruise it sank in Mobile Bay during a storm.

Undaunted, Hunley started work on his fourth undersea warship, and he named it *H.L. Hunley*. Roughly four feet in diameter, the iron submarine required a crew of eight men, one to navigate and steer, and seven to turn the hand-cranked propeller shaft. Each end of the 37-foot-long submarine featured ballast tanks that could be flooded, or pumped dry by hand pumps. Iron weights bolted to the underside of the hull provided extra ballast. In an emergency this iron ballast could be removed by unscrewing heads of bolts inside the vessel. *Hunley* featured two watertight hatches for ingress and egress, one forward and one aft.

Armament consisted of a single spar torpedo, a copper cylinder containing 90 pounds of black powder. The torpedo was mounted on a long 23-foot wooden spar that extended forward from the bow. This torpedo had a barbed metal tip. In an attack *Hunley* would ram the torpedo into a targeted wooden warship. Impact would drive the barbed tip firmly into the ship's hull, where it would remain. *Hunley* then would back away, and the torpedo would be detonated by pulling an attached rope lanyard.

Hunley arrived in Charleston via rail on 12 August 1863. She never was commissioned into service with the C.S. Navy. Instead she would operate under control of the C.S. Army in coordination with the Confederate Secret Service. On 29 August she made her first test run in Charleston Harbor. At first all went well as *Hunley* cruised on the surface with her two hatches open. Then her captain, Lt. John Payne, chanced to step on the diving planes lever. The vessel immediately dove underwater with her hatches still open, and she flooded within seconds. Lt. Payne and two other men managed to clamber out, but the remaining five volunteer C.S. Army crewmen drowned.

The army raised and refitted *Hunley* and prepared for another harbor test. On 15 October, crewed by eight more volunteers, the submarine submerged and made a successful mock attack. Yet, after the torpedo detonated *Hunley* failed to resurface. All eight crewmen – including Horace Hunley himself, who was aboard – were lost.

Once again the Army raised and refitted *Hunley*. With eight more volunteer crewmen she crept out of Charleston Harbor on the night of 17 February 1864. Somewhere in the darkness beyond the harbor entrance she submerged and headed for anchored Union warships. *Hunley* set her sights on the 1,260-ton *USS Housatonic*, a steam sloop-of-war five miles offshore. *Hunley's* barbed spar torpedo jammed into the warship hull below the waterline. In a few moments the torpedo detonated and blasted a giant hole in the wooden hull. *Housatonic* sank quickly.

After the successful attack *Hunley* vanished. She never returned to her dock in Charleston Harbor. Over passing years many credible loss scenarios have been postulated, but the cause never has been positively established. Nonetheless, *Hunley* earned her niche in military history. She had become the first submarine to sink an enemy warship.

Over 100 years later the sunken *Hunley* would be found under 27 feet of water on the seaward side of the *Housatonic* wreckage site. Plans to raise the famous submarine would span five years. Pulled from the seabed by the recovery barge *Karlissa B.,* the *Hunley* once again saw daylight on 8 August 2000. Ecstatic crowds in surrounding boats and on the shore greeted her resurrection from a watery grave.

The recovery barge brought *Hunley* back to Charleston. Remains of her crew were still aboard and still seated at their assigned stations. This discovery added fuel to the many loss scenarios that had been suggested over the years. After several years of evaluation, preservation, and restoration the *Hunley* could be viewed by the public at Warren Lasch Conservation Center in Charleston. A replica is displayed at Battleship Memorial Park in Mobile, Alabama.

CSS Alabama (aka, *Enrica*): James Bulloch, the Confederate agent in Liverpool, arranged for construction of a warship in the English shipyard at Birkenhead. To conform with British neutrality laws the vessel would be designed as a warship for the Confederacy, but it would not be armed until it reached international waters. The new ship featured reinforced decks for canon placement, powder magazines below the waterline, etc. She launched as the civil *Enrica* in May 1862. Bulloch arranged for a civilian crew to sail the ship to Terceira Island in the Azores.

In the Azores the weaponless new sloop-of-war met the civil steamers *Bahama* and *Agrippina*. They brought the new Confederate commander, Capt. Raphael Semmes, and his crew. They also brought a full supply of coal plus cannons, shot, powder, etc. Within a week the guns had been mounted. The ship re-coaled and took aboard necessary supplies. By 24 August 1862 the conversion from civil vessel to warship had been completed. Capt. Semmes re-commissioned *Enrica* as the new naval cruiser *CSS Alabama*.

The warship displaced 1,050 tons and measured 220 feet in length. In addition to her sails she also had two steam engines and carried 350 tons of coal. "Lifting gear" could retract her screw propeller to reduce drag when using only sail power. She could make 10 knots under sail power alone and 14 knots when using both sail and steam. The fully rigged triple-masted vessel had been designed for a crew of 145 officers and men, including the gun crews. Her ordnance consisted of six 32-pounder smoothbore guns plus two much heavier 100-pounder long range guns, one of which was rifled. The motto, "God helps those who help themselves," was engraved in bronze on the ship's wheel.

CSS Alabama was destined to be a commerce raider. Capt. Semmes had brought only a skeleton crew with him, but he enlisted 83 more men from crews of the *Agrippina* and the *Bahama*. Semmes spent two months in the eastern Atlantic, capturing and burning Union merchant ships. Then he sailed across the Atlantic and into the West Indies where the hunting improved. Semmes met a Union warship, *USS Hatteras*, along the Texas coast near Galveston. In his first battle with an enemy warship he quickly sank the sidewheel steamer. Then he continued south and plundered Union merchant shipping along the Brazilian coast.

Alabama again crossed the Atlantic and stopped for provisions and refitting at Cape Town near the southern tip of Africa. Thereafter she sailed east and spent six months preying on Union shipping in the East Indies. Captured crews were not harmed. As soon as practical they were placed aboard neutral ships or put ashore at neutral ports. Thus far no Confederate crewman had died, and *Alabama* had never entered a Confederate port. Union warships hunted the now-famous raider. Yet, the few times she was sighted she outran or outwitted her pursuers and vanished into the dark or over the horizon.

Capt. Semmes returned to European waters and dropped anchor in the neutral harbor at Cherbourg, France, on 11 June 1864. He wanted to use the dry dock there to overhaul his ship after spending so much time at sea. During 534 days of commerce raiding his crew had boarded about 440 ships, had burned 65 of them, and had captured over 2,000 prisoners. In terms of the number of Union merchant ships destroyed, *Alabama* had been the most destructive commerce raider during the entire war.

However, trouble loomed ahead for *Alabama* and her crew. Three days after she entered the harbor a big Union warship, *USS Kearsarge*, arrived and anchored outside the harbor entrance. *Alabama* was trapped. Worse yet, Capt. Semmes realized that more enemy warships now would be on the way. For the moment he was safe from Union attack inside the harbor of a neutral country. However, one historian explained Semmes' reasoning:

> Having no desire to see his worn-out ship rot away at a French dock while quarantined by Union warships, and given his instinctive aggressiveness and a long held desire to once again engage his enemy, Capt. Semmes chose to fight.

On 19 June 1864 the *Alabama* steamed out of the harbor to fight *Kearsarge*. Once the ships reached international waters the deadly dance began. The two warships steamed in circles on opposite headings, each trying to cross the bow of her opponent (called, *crossing the T*) in order to fire a more effective broadside. Both vessels suffered hits, but *Alabama* fared far worse. At 1,550 tons, *Kearsarge* was 50 percent bigger than *Alabama*, and she had bigger guns. Worse yet, unknown to Capt. Semmes, *Kearsarge* had iron chain armor amidships from her decks to the waterline. Further, survivors from both sides later would explain that accuracy of fire from *Kearsarge* was superior to fire from *Alabama*.

Massive 11-inch guns aboard *Kearsarge* took a lethal toll.

After about an hour of battle *Alabama* was ravaged, and she began to sink. Capt. Semmes struck his colors. As *Alabama* slid under the waves, *Kearsarge* picked up many survivors. However, Semmes and 41 members of his crew were pulled from the ocean by the British civil steamer *Deerhound*. Thus, they were not captives and went to England as free men.

One-hundred-twenty years later in 1984 the French Navy ship *Circe* would discover the sunken remains of *Alabama*. She lay under 200 feet of water east of Cherbourg. In 1988 the CSS Alabama Association began scientific exploration of the wreck. By the year 2002, divers would recover over 300 artifacts including the ship's bell, her cannons, tableware, commodes, and other items. They would provide invaluable insight into life aboard the famous *CSS Alabama*.

CSS Virginia (aka, *Merrimack*): Opening days of the war led to birth of the most famous Confederate warship. Prior to hostilities the huge Gosport Navy Yard (later, *Norfolk Navy Shipyard*) was one of the most important Union military bases. When Virginia seceded on 17 April 1861 the Union tried to destroy the base so it would not fall into Confederate hands. As part of the destruction the U.S. Navy burned and sank their big steam frigate, *USS Merrimack*.

Within days the C.S. Navy occupied the naval base. The new base commander, Adm. French Forrest, found that most base provisions and munitions had escaped Union efforts to destroy them. *Merrimack* had burned to the waterline and had sunk, but Forrest believed she could be salvaged. Although submerged, her intact lower hull plus her steam engine and machinery had not been materially damaged.

Salvage work began in July 1861. Workmen raised the hull and towed it into a dry dock. They cut the ship's burned timbers and superstructure down past the original waterline, and then a conversion began. Using the gutted lower hull, engine, and machinery, the former wooden Union frigate would become a new Confederate casemate ironclad.

At 275 feet in length, the massive new warship would displace about 4,100 tons. She would be powered by two horizontal steam engines fired by four boilers. Due to her great weight she had an extremely deep draft of 22 feet and could manage a top speed of only five or six knots. Workmen built a new armored casemate atop a new armored deck. The thick casemate would be impervious to enemy fire. Layered oak and pine sheeting, two feet thick, would be covered by four inches of iron armor. This defensive bulwark would be angled 54 degrees from the vertical to deflect smoothbore and rifled enemy shells.

The casemate featured 14 gunports, three in the bow, three in the stern, and four on each side. All were protected by iron-covered shutters. The ship's biggest guns were her six 9-inch Dahlgrens weighing 14,500 pounds each. They had been salvaged from the burned hull. Two of them near the boilers could fire heated shot. In addition there were four muzzle-loading 7-inch Brooke rifles. Further, the ship carried two 7-inch pivot guns, one each in the bow and stern, plus two 6.4-inch cannons. Above the casemate on the deck stood two 12-pounder howitzers, one forward and one aft. They could be loaded with cannister to thwart boarders. A massive iron ram had been fitted onto the ship's bow.

The new warship was commissioned *CSS Virginia*, and Capt. Franklin Buchanan took command. From a firepower and defensive standpoint his ship was indeed a formidable behemoth, but she did have vulnerabilities. She was slow, cumbersome, and had a turning radius of about one mile. Her freeboard was negligible, so she was not suited for steaming on the open sea. She would be limited to service in coastal bays and harbors.

On 8 March 1862, *CSS Virginia* sortied into Hampton Roads, a four-mile-wide channel through which the James, Elizabeth, and Nansemond Rivers flow into the Chesapeake Bay. She fired upon blockader *USS Cumberland* and then rammed her. As *Cumberland* sank the Confederate ironclad turned toward *USS Congress*. Seeing what had happened to the first victim, *Congress* tried to flee but ran aground. The Confederate ironclad remained in deep water and shelled the grounded ship. The cannonade reduced the Union man-of-war to a burning wreck. She surrendered and later sank.

Virginia turned toward the *USS Minnesota*, which ran aground on a sandbar in an effort to get away. In the shallow coastal water *Virginia* dared not get close enough to destroy the stranded Union ship. *Virginia* set her sights on the warships *USS Roanoke* and *USS St. Lawrence*. Both of them ran aground, perhaps intentionally, in an effort to escape from the ironclad's ram. Late afternoon arrived and the tide was falling, so Capt. Buchanan elected to retire. He planned to return the next morning and complete the destruction of blockading Union ships.

Already the U.S. Navy had suffered a crushing loss. Two men-of-war had been destroyed, three more had run aground, and over 400 sailors had been killed. That night Lincoln's cabinet met in Washington. Edward Stanton, the panicky Secretary of War, said "the rebel monster" could steam up the seacoast and shell coastal cities. He added that *Virginia* also might cruise up the Potomac River and shell Washington (in reality *Virginia* dared not venture out onto the open sea, and the Potomac was too shallow for her 22-foot draft).

That night a new Union ironclad, *USS Monitor*, was towed into Hampton Roads. The next morning on 9 March the world would witness the first clash between two ironclad warships. The much smaller and nimbler *Monitor* featured a rotating turret with two 11-inch guns. She could outmaneuver *Virginia*, but the Confederate ironclad had the advantage in firepower. Both ships closed and blasted away at each other, often at point-blank range. After several hours of combat no crippling damage had been done to either warship.

Yet, numerous hits on *Monitor's* pilothouse forced the crew to assess the damage. In addition, while looking through viewing slits her captain, Lt. John Worden, had been blinded by *Virginia's* explosive shells. *Monitor* suddenly withdrew into shallow water where *Virginia* could not follow. Unable to get close enough to attack her opponent without fear of running aground, *Virginia* waited for an hour. Then she withdrew and headed back to her James River dock.

The next month on 11 April the *CSS Virginia* once again

steamed into Hampton Roads, but Union ships remained in shallow water and obeyed orders not to engage. Crewmen from *Virginia* and her supporting ships boarded and seized three Union merchant ships anchored in the harbor: *Marcus*, *Sabout*, and *Catherine T. Dix*. Nonetheless, Union warships remained in shallow water and refused to steam out and fight *Virginia*. Ironically, neither *Virginia* nor *Monitor* ever would fight anyone again.

Attacking Union Army troops occupied Norfolk on 10 May 1862. *Virginia* could not steam out to sea and escape, for she had not been designed for the open ocean. She also could not retreat up the James River because of her deep 22-foot draft. She was trapped. To prevent her capture her crew set her afire. When flames reached her magazines the tooth-rattling blast could be heard 35 miles away.

Monitor met her fate on New Year's Eve that same year. Like *Virginia*, she could not cruise on the open ocean and was being towed southward by the tug *USS Rhode Island*, bound for the federal blockade at Wilmington. However, ocean swells swamped *Monitor*, and she sank off Cape Hatteras, North Carolina. One-hundred-eleven years later in 1973 her sunken wreckage would be discovered 16 miles offshore.

CSS Virginia and *USS Monitor* had ushered in a new era of war at sea. All navies in the world learned an invaluable lesson. Thereafter ironclad ships, and later ships with solid iron or steel hulls and superstructures, would become the worldwide military standard for nautical men-of-war.

List of C.S. Navy Ships: There is no *official* listing of C.S. Navy ships. Some military vessels were constructed domestically by the Confederacy during the war, while others were built in European shipyards. Some ships that found their way into the C.S. Navy were converted merchantmen that had been captured on the high seas. Others were captured U.S. Navy warships that had been docked or anchored in southern ports when the war began.

Some ships that at one time or another flew the Confederate flag were ocean-going vessels, but most others were limited to harbor or river defense. Some of these ships were powered by steam, while others were powered by sail alone. A few floating batteries had no propulsion whatsoever and had to be towed into firing position. A few C.S. Navy ships survived the war, but most did not.

The following list is not all-inclusive. It has been compiled from various sources. This list excludes the many hundreds of privately owned blockade runners and privateers:

C.S. Navy Cruisers: Cruisers were ocean-going ships. Most were armed commerce raiders. Typically they were lightly armed and very fast wooden steamers, although a few were powered by sail. The Confederacy planned to add ironclad cruisers to her fleet, and three such vessels were under construction in Europe. However, only one of them, *CSS Stonewall*, was completed and commissioned. It arrived too late to have an impact during the war:

– *CSS Alabama*: screw steamer, commerce raider, sunk 1864.
– *CSS Alexandra*: screw steamer, seized before delivery, 1863.
– *CSS Archer*: schooner, commerce raider, captured 1863.
– *CSS Chickamauga*: screw steamer, commerce raider, burned 1865.
– *CSS Clarence*: brig, commerce raider, burned 1863.
– *CSS Florida*: screw steamer, commerce raider, captured 1864.
– *CSS Georgia*: screw steamer, commerce raider, sold 1864.
– *CSS Georgiana*: steamer, destroyed after leaving port, 1863.
– *CSS Lapwing*: bark, burned 1863.
– *CSS Nashville*: sidewheel steamer, commerce raider, sunk 1862.
– *CSS Rappahannock*: steamer, sloop-of-war, survived the war.
– *CSS Shenandoah*: steamer, commerce raider, survived the war.
– *CSS Sumter*: screw steamer, commerce raider, sold 1862.
– *CSS Tacony*: bark, commerce raider, burned 1863.
– *CSS Tallahassee*: screw steamer, commerce raider, seized 1865.
– *CSS Tuscaloosa*: bark, commerce raider, seized 1863.
– *CSS United States*: sail frigate, harbor defense, scuttled 1862.
– *CSS North Carolina*: ironclad, seized before completion, 1863.
– *CSS Mississippi*: ironclad, seized before completion, 1863.
– *CSS Stonewall*: ironclad, survived the war.

C.S. Navy Ironclad Steamers: The C.S. Navy wanted heavily armored ships to defend its vital rivers and harbors and keep them open for commerce. To navigate winding rivers or confined harbors, such ships could not rely on sail power alone. Steam propulsion was

essential. To battle invading U.S. Navy warships, ironclad vessels would be required. The C.S. Navy built such vessels in shipyards along major rivers and harbors in the South. These ships were not designed for service on the open ocean:
– *CSS Albermarle*: ironclad twin-screw steamer, sunk 1864.
– *CSS Arkansas*: ironclad ram twin-screw steamer, scuttled 1862.
– *CSS Atlanta*: ironclad triple-screw steamer, captured 1863.
– *CSS Baltic*: ironclad ram steamer, captured 1865.
– *CSS Charleston*: ironclad steamer, destroyed 1865.
– *CSS Chicora*: ironclad ram screw steamer, destroyed 1865.
– *CSS Columbia*: ironclad ram screw steamer, captured 1865.
– *CSS Eastport*: ironclad steamer under construction, captured 1862.
– *CSS Fredericksburg*: ironclad twin-screw steamer, destroyed 1865.
– *CSS Huntsville*: ironclad steam battery, scuttled 1865.
– *CSS Louisiana*: ironclad twin-screw steamer, destroyed 1862.
– *CSS Manassas*: ironclad ram screw steamer, sunk 1862.
– *CSS Milledgeville*: ironclad steamer, burned and sunk 1864.
– *CSS Missouri*: ironclad steam sloop, surrendered 1865.
– *CSS Mobile*: ironclad screw steamer, burned before launch 1863.
– *CSS Muscogee*: ironclad twin-screw steamer, burned 1865.
– *CSS Nashville*: ironclad sidewheel steamer, surrendered 1865.
– *CSS Neuse*: ironclad ram twin-screw steamer, destroyed 1865.
– *CSS New Orleans*: ironclad floating steam battery, scuttled 1862.
– *CSS Palmetto State*: ironclad ram sloop, destroyed 1865.
– *CSS Raleigh*: ironclad sloop, ran aground and destroyed 1864.
– *CSS Richmond*: ironclad screw steamer, scuttled 1865.
– *CSS Savannah*: ironclad steam sloop, burned 1864.
– *CSS Tennessee* (1): ironclad ram steamer, destroyed 1862.
– *CSS Tennessee* (2): ironclad ram screw steamer, captured 1864.
– *CSS Texas*: ironclad ram screw steamer, captured 1865.
– *CSS Tuscaloosa*: ironclad ram steamer, scuttled 1865.
– *CSS Virginia*: ironclad ram screw steamer, scuttled 1862.
– *CSS Virginia II*: ironclad ram screw steamer, destroyed 1865.
– *CSS Wilmington*: ironclad twin-screw steamer, destroyed 1865.

<u>C.S. Navy Cotton-Clad Steamers</u>: To defend inland rivers from attack and patrol by Union gunboats, the Confederacy needed vessels that were more easily built than huge and expensive ironclads. Also, such vessels did not need to be as heavily armed and armored as ships-of-

the-line. For this purpose the Confederacy turned to cotton-clad ships. Such ships typically had one-inch-thick iron armor covering their bow, and some also had iron rams. However, the rest of the ship was constructed with pine or oak double wooden bulkheads. In the wide hollow space between bulkheads, builders inserted compressed cotton bales. Enemy fire could easily penetrate the outer wooden bulkhead. Yet, compressed cotton bales between bulkheads would absorb the blow, and the inner bulkhead would remain intact. Some of these cotton-clad military steamers are listed below:
– *CSS Colonel Lovell*: cotton-clad sidewheel steamer, sunk 1862.
– *CSS General Beauregard*: cotton-clad ram steamer, sunk 1862.
– *CSS General Bragg*: cotton-clad ram steamer, captured 1862.
– *CSS Breckinridge*: cotton-clad ram steamer, burned 1862.
– *CSS Defiance*: cotton-clad sidewheel ram steamer, burned 1862.
– *CSS Gen. Earl Van Dorn*: cotton-clad ram steamer, burned 1862.
– *CSS Gen. Jeff Thompson*: cotton-clad ram steamer, sunk 1862.
– *CSS Gen. Sterling Price*: cotton-clad ram steamer, sunk 1862.
– *CSS Gen. Sumter*: cotton-clad ram steamer, captured 1862.
– *CSS Governor Moore*: cotton-clad ram steamer, destroyed 1862.
– *CSS Little Rebel*: cotton-clad ram steamer, captured 1862.
– *CSS Resolute*: cotton-clad ram sidewheel steamer, captured 1864.
– *CSS Stonewall Jackson*: cotton-clad ram steamer, burned 1862.
– *CSS Warrior*: cotton-clad sidewheel ram steamer, destroyed 1862.
(the following ships were operated by the C.S. Army)
– *CSS Grand Duke*: cotton-clad steamer, burned 1863.
– *CSS Josiah A. Bell*: cotton-clad steamer in Texas, scuttled 1865.
– *CSS Queen of the West*: cotton-clad ram steamer, exploded 1863.
– *CSS Uncle Ben*: cotton-clad steamer in Texas (fate unknown).
– *CSS Webb*: cotton-clad ram steamer, burned 1865.

C.S. Navy Gunboats: The term "gunboat" embraces different sizes and types of military ships. Almost all were powered by steam, were lightly armed and armored, and were used for river defense. Most had been civil vessels before the war and were hastily armed after hostilities began:
– *CSS Appomattox*: converted tugboat, burned 1862.
– *CSS Bayou City*: surrendered 1865.
– *CSS Beaufort*: screw steamer, captured 1865.
– *CSS Bienville*: sidewheel steamer, destroyed 1862.

– *CSS Black Warrior*: schooner, burned 1862.
– *CSS Bombshell*: steamer, captured 1864.
– *CSS Calhoun*: sidewheel steamer, captured 1862.
– *CSS Carondelet*: sidewheel steamer, destroyed 1862.
– *CSS Chattahoochee*: screw steamer, scuttled 1864.
– *CSS Clifton*: sidewheel steamer, scuttled 1864.
– *CSS Curlew*: sidewheel steamer, sunk 1862.
– *CSS De Soto*: sidewheel steamer, captured 1862.
– *CSS Diana*: steamer, survived the war.
– *CSS Drewry*: steamer tender, destroyed 1865.
– *CSS Ellis*: steamer tugboat, captured 1862.
– *CSS Equator*: steamer, burned 1865.
– *CSS Fanny*: iron-hulled screw steamer, burned 1862.
– *CSS Fashion*: schooner (fate unknown).
– *CSS Forrest*: steamer tugboat, burned 1862.
– *CSS Gaines*: sidewheel steamer, sunk 1864.
– *CSS General Quitman*: steamer, destroyed 1862.
– *CSS George Polk*: sidewheel steamer, destroyed 1862.
– *CSS General Page*: sidewheel steamer, burned (date uncertain).
– *CSS Germantown*: sloop-of-war, sunk as blockship 1862.
– *CSS Governor Moore*: sidewheel steamer, destroyed 1862.
– *CSS Hampton*: screw steamer tug, scuttled 1865.
– *CSS Helen*: sidewheel steamer, harbor gunboat, sunk 1864.
– *CSS Henry Dodge*: cutter, schooner rigged (fate unknown).
– *CSS Huntress*: sidewheel steamer, burned 1863.
– *CSS Isondiga*: steamer, burned 1864.
– *CSS Ivy*: sidewheel river steamer, burned 1863.
– *CSS J.A. Cotton*: sidewheel river steamer, burned 1863.
– *CSS Jackson*: sidewheel river steamer, destroyed 1862.
– *CSS Jamestown*: sidewheel steamer, sunk 1862.
– *CSS Junaluska*: tug steamer, scrapped 1862.
– *CSS Kate Bruce*: schooner, scuttled (date unknown).
– *CSS Lady Davis*: steamer tug, scrapped 1865.
– *CSS Livingston*: sidewheel steamer, destroyed 1862.
– *CSS Macon*: steamer, survived the war.
– *CSS Maurepas*: sidewheel steamer, sunk 1862.
– *CSS McRae*: screw steamer, sloop rigged, sunk 1862.
– *CSS Morgan*: sidewheel steamer, surrendered 1865.

– *CSS Morning Light*: sail rigged, burned 1863.
– *CSS Nansemond*: twin screw gunboat, burned 1865.
– *CSS Neptune*: steamer, sunk 1863.
– *CSS Oregon*: steamer, scuttled 1862.
– *CSS Pamlico*: sidewheel river steamer, burned 1862.
– *CSS Patrick Henry*: sidewheel steamer, burned 1865.
– *CSS Pee Dee*: screw steamer, river gunboat, sunk 1865.
– *CSS Pickens*: cutter, schooner rigged (fate unknown).
– *CSS Plymouth*: sloop-of-war, burned 1862.
– *CSS Pontchartrain*: sidewheel river steamer, burned 1863.
– *CSS Rappahannock*: sidewheel steamer, burned 1862.
– *CSS Roanoke*: screw steamer, destroyed 1865.
– *CSS Savannah*: steamer, floundered and sank 1863.
– *CSS Schultz*: sidewheel river steamer, sunk 1865.
– *CSS Sea Bird*: sidewheel river steamer, sunk 1862.
– *CSS Selma*: sidewheel river steamer, captured 1864.
– *CSS Talomico*: sidewheel steamer, sunk 1863.
– *CSS Teaser*: tug steamer, captured 1862.
– *CSS Torpedo*: screw steamer, burned 1865.
– *CSS Water Witch*: sidewheel steamer, burned 1864.
– *CSS Yadkin*: steamer, burned 1865.

C.S. Navy Blockade Runners: During the war many hundreds of ships were engaged in running the coastal blockade. Typically they were very fast wooden steamers. Most were privately owned. However, a few were ships of the C.S. Navy, and they are listed below:
– *CSS Advance*: sidewheel steamer, captured 1864.
– *CSS America*: racing yacht, scuttled 1862.
– *CSS Harriet Lane*: sidewheel steamer, captured 1863.
– *CSS Kate Dale*: sidewheel steamer, captured 1863.
– *CSS Lady Stirling*: sidewheel steamer, captured 1864.
– *CSS Owl*: steamer, survived the war.
– *CSS Rob Roy*: steamer, survived the war.
– *CSS Robert E. Lee*: iron-hulled steamer, captured 1863.
– *CSS William G. Hewes*: sidewheel steamer, captured 1863.
– *CSS Sumter*: screw steamer, bark rigged, sold 1862.

Submarines and Semi-Submersibles: Vessels designed for underwater or semi-submerged service were not a novel innovation during the war. They had been in civil use for underwater salvage purposes for several years. However, various experiments with military usage began in earnest during the war.

During that era, marine explosives were known as "torpedoes." Vessels designed to carry and deploy such explosives were called "torpedo boats." During the war such vessels featured revolutionary experimental technology. Some became part of the C.S. Navy, while others operated under the aegis of the Confederate Secret Service. Some of these vessels are listed below:

– *CSS David (1)*: semi-submersible torpedo boat, abandoned 1863.
– *CSS David (2)*: semi-submersible torpedo boat, captured 1865.
– *CSS Midge*: steam torpedo boat, captured 1865.
– *CSS Saint Patrick*: experimental torpedo boat, abandoned 1864.
– *CSS Squib*: spar torpedo boat (fate unknown).
– *CSS Hornet*: steam spar torpedo boat, sank 1865.
– *CSS Scorpion*: steam spar torpedo boat, abandoned 1865.
– *CSS Wasp*: steam spar torpedo boat (fate unknown).
(the following were privately owned torpedo boats)
– *H.L. Hunley*: submersible torpedo boat, vanished at sea 1864.
– *Pioneer*: experimental submersible, scuttled 1862.
– *Pioneer II*: experimental submersible, abandoned 1863.
– *American Diver*: experimental submersible, sank 1863.

<u>Others Ships</u>: Many other vessels took part in the four-year war, and most were privately owned. Hundreds of privately owned ships engaged in blockade running and tried to avoid contact with U.S. Navy warships and U.S. Army shore batteries. In addition scores of privateers sailed as quasi-pirates during the war.

There were 19 known C.S. Navy transports, tenders, and tugs. The Confederacy also had one known hospital ship, *CSS Kanawha Valley*, a stern-wheel river steamer. Relatively little has been written about nautical aspects of the American Civil War, but maritime forces in the North and South played a significant role.

Fanciful etching of Sir Francis Drake, sometimes a pirate, and at other times a privateer, looking at captured Spanish treasure.

CSS Alabama, a sail and steam commerce raider. She was sunk in 1864 by USS Kearsarge near Cherbourg, France.

<u>CSS Stonewall</u>, a keel-up ironclad behemoth built in France for the C.S. Navy. She arrived in the Caribbean after the war ended.

<u>CSS Shenandoah</u>, a highly effective sail and steam commerce raider. She survived the war, circumnavigated the globe, sailed to England, and surrendered seven months after the war ended.

CSS Virginia (aka, Merrimack), the first ironclad warship in the world to go into battle. She was scuttled and burned in 1862.

CSS Atlanta, a casemate ironclad similar to CSS Virginia. This heavily armored triple-screw steamer was captured in 1863.

Etching of H.L. Hunley, a submersible torpedo boat. She was the first submarine in history to sink an enemy warship. However, after she sank USS Housatonic, she sank also, and her crew was lost.

CSS Robert E. Lee, a schooner rigged, iron hulled, paddle steamer. As a blockade runner, she ran aground and was captured in 1863.

CSS Manassas, a small cigar-shaped ironclad ram steamer. She was used on the lower Mississippi River and was sunk in 1862.

CONFEDERATE STATES STEAMER SUMTER.

CSS Sumter, a bark-rigged steam screw steamer used as a commerce raider by the C.S. Navy. She was captured while in port at Gibraltar in December 1862 and sold to the British government. Later when privately owned and under British registry, she became the speedy blockade runner Gibraltar of Liverpool.

Seal of the Confederate States Navy Department.

Lo, all our pomp of yesterday
Is one with Ninevah and Tyre!
Judge of the Nations, spare us yet,
Lest we forget – Lest we forget!
[Rudyard Kipling, Recessional, 1897]

Almost Home – The Sultana Disaster

> I did not intend to be killed today.
> *[Vicomte de Turenne, a mortally wounded French soldier, as he lay dying after the Battle of Salzbach, 27 July 1678]*

Setting the Stage: Richmond, the Confederate capital city, had been evacuated by the C.S. Government and C.S. Army. A week later on 9 April 1865 the Army of Northern Virginia surrendered. Other southern commands soon would follow suit. Jefferson Davis, the Confederate president, and his civil and military entourage would be captured in rural Georgia on 10 May. For all practical purposes the Confederate States of America would cease to exist.

Prisoners of war in the North and South rejoiced. Many of them were semi-invalids, half-starved and suffering from a host of diseases and ailments because of contaminated water, overcrowding, lack of medical care, lack of food, and exposure. Many of their comrades had died in the squalid POW camps. Yet, survivors knew that they would be going home, and they were impatient to get started. Once they reached home they hoped to forget the horrors of prison camp life. They wanted to resume life with their families and get the rest, good food, love, and comfort they badly needed.

On the eastern bank of the mighty Mississippi River, Vicksburg, Mississippi, had become a repatriation hub for former Union POWs. The U.S. Army began contracting with steamboat captains who could ferry former POWs upriver to army camps in the North.

The Steamboat: Many newly released prisoners were destined to board the *Sultana*, a 260-foot-long sidewheel river steamboat. Capt. Cass Mason commanded the big steamer. It had been built during 1863 at the John Litherbury Boatyard in Cincinnati. Listed at 1,719 tons, the craft was designed to carry passengers, livestock, cotton, and other cargo. She was operated by 85 crewmembers. By law her maximum capacity was 376 persons, including her crew. During the past two years the steamboat had shuttled cargo and passengers on the lower Mississippi between New Orleans, St. Louis, and ports in between. U.S. War Department records confirm that Union troops had been a major source of passenger revenue during the war.

On 21 April 1865, *Sultana* left New Orleans and headed upriver with a hundred hogsheads of sugar, 96 assorted heads of livestock, and 87 paying passengers. Three days later on the evening of 24 April she made a scheduled stop at Vicksburg. While moored at the dock, Capt. Mason discovered that one of his four boilers was leaking. His maintenance engineers recommended drawing fires, cutting out the ruptured seam, and replacing it. Mason knew that process would take several days, but he was in a hurry. He and his chief engineer decided to hammer the faulty seam back into place and rivet a metal patch over the ruptured area.

While repairs were underway Mason got a visit from LtCol. Reuben Hatch, the U.S. Army quartermaster at Vicksburg. Hatch was in charge of thousands of former Union POWs who had been released from prison camps at Andersonville, Georgia, and Cahaba, Alabama. These men were waiting to be carried north and repatriated. Hatch agreed to pay Mason for carrying some of the ex-prisoners upriver to Cairo, Illinois. At Cairo the repatriated soldiers were to get detailed medical examinations and treatment. Then they would be discharged from the U.S. Army, given back pay, loaded onto trains, and sent back to their hometowns.

Former prisoners in question originally hailed from the states of Ohio, Michigan, Indiana, Kentucky, and West Virginia. Hatch agreed to pay five dollars for each enlisted man and ten dollars for each officer that Mason would carry to Cairo, at the confluence of the Mississippi and Ohio Rivers. For Mason this looked like a golden opportunity. The two men agreed that about 1,400 former POWs could be carried.

Fate intervened with a mixup in parole camp rosters. Also, other steamboat captains were eager to reap some of the easy money by carrying former POWs to Cairo. A U.S. Army officer, Capt. George Williams, was in charge at the Vicksburg parole camp. He somehow understood that he was to send *all* paroled prisoners to *Sultana*.

The River Voyage: Former prisoners began boarding the steamboat. Most were still weak from months of captivity, but they were a happy lot. One observer on the dock described them as "the most lighthearted a crowd as ever came up a gangplank." On and on they came by the hundreds, far eclipsing the legal limit of 376 souls. Faced with the masses coming aboard, Capt. Mason elected to forego making out the passenger manifest at Vicksburg. He planned to take care of that formality once his craft was underway, for he had been guaranteed a handsome profit. One historian explained:

> Altogether there were probably some 2,300 persons on the steamer when the lines were cast off. Naturally the boat was almost unbelievably crowded. The soldiers were *[standing]* on the hurricane deck; all available space was filled. They packed the steamer from top to bottom, cabins, Texas deck, even the pilothouse. Almost literally, the steamer could not have carried another human being.

Sultana left Vicksburg and began puffing its way upriver. Capt. Mason was worried. He was fighting a current made much stronger than normal by the swollen river, which was in springtime flood stage. Nonetheless the steamboat made several scheduled stops. At 7:00 p.m. on the evening of 26 April it eased to the dock at Memphis, Tennessee. Stevedores began unloading the hogsheads of sugar that weighed about 115 tons.

Before midnight after the sugar had been unloaded, the crew slipped the mooring lines. *Sultana* crossed to the western bank of the river and took on more coal. Then the steamer headed upriver into the night. The next scheduled stop was at Cairo, where the former POWs would debark. One of the ship engineers later would note:

> The current was strong, and Sultana was fearfully overloaded, with six times as many passengers as she had been designed to carry.

Midnight came and passed. Progress against the strong current was slow, and the steamboat was heavily laden. The paddlewheels clawed at the water, pulling the craft up the river. By 2:00 a.m. the soldiers and most crewmen were asleep. Men were curled up on the decks everywhere as the steamboat mushed past river islands known as The Chickens and Hen about seven miles north of Memphis.

The Explosion and Fire: Deep in the ship's bowels heavy steam pressure finally exceeded the strength of the repaired boiler. It exploded with a roar akin to a thunderclap. Within seconds two more weakened boilers exploded as well. One narrative of the events explains the result:

> Hundreds of sleeping soldiers were blown into the river. *[They had been]* snugly asleep one moment, and hurled through the air and into the cold black water the next. With them went great chunks of twisted machinery, a shower of red-hot coals, . . . wood, cabin furniture, railings and deck beams. Half of the steamboat had disintegrated.

A blazing inferno lit up the sky for miles in all directions. The deafening noise of exploding boilers had been heard in Memphis. Those awakened by the blasts could see the orange glow from flames north of the river city. *USS Grosbeak*, a Union gunboat docked at Memphis, cast off her lines and pounded upriver at full speed. Civilian craft plying the river that night also heard the explosions and saw the sky turn orange. They all headed toward the glow to provide whatever aid they could offer. The southbound steamer *Bostona II* was on her maiden voyage, and she reached the burning wreckage within the hour. Other steamers soon arrived: *Silver Spray*, *Jenny Line*, *Pocahontas*, and U.S. Navy gunboats *Essex* and *Tyler*.

Sultana had become a blazing wreck drifting downstream. The deck supporting passenger cabins had collapsed, dropping occupants down into raging fires below. The two tall smokestacks had come crashing down, pinning and trapping many men. The blasts had scattered burning coal from the furnaces, turning the midships section into a fiery cauldron.

Men who had not been blown overboard into the river still clung to parts of the bow and stern. Flames soon gave them a choice: stay on the wreckage and be consumed by the fire, or jump into the dark water. Most chose the latter option.

The water was cold. Many men of that era had never learned how to swim. Others who could swim had been weakened by months of captivity in a prison camp where they had been almost starved. When the steamboat boilers exploded some men had been scalded by escaping steam, injured by the blasts, or horribly burned. In the water there was little wreckage to which they might cling. One survivor later would write about the catastrophe:

> I got about 300 yards from the boat, clinging to a heavy plank. The whole heavens seemed to be lighted up by the conflagration. Hundreds of my comrades were *[crushed in the wreckage]* by the heavy timbers of the upper decks, and they burned. The water seemed to be one solid mass of human beings struggling with the river.

Another survivor would write about men jumping into the water and then searching for something to which they might cling:

> I would see men jumping from all parts of the boat into the water until it seemed black with men, their heads bobbing up and down like corks, and then disappearing beneath the turbulent waters, never to appear again.

Several hundred injured, burned, or scalded former prisoners still huddled on the bow of the boat, for it would be the last section to burn. They screamed for help. Many prayed for divine intervention. As flames drew ever closer to them, many of these men tore off doors and timbers and threw them overboard. Then they jumped in after them. Anything was better than waiting for the advancing flames. However, a few men refused to leap into the river. One survivor explained the fate of men who tried to stay on the burning wreckage:

> The men who were afraid to take to the water could be seen clinging to the bow of the boat until they were singed off like flies *[by the flames]*.

Wreckage floated six miles downstream, struck an island in the river near Marion, Arkansas, and lodged there. Remnants of what once had been a steamship slowly sank. Pieces of wreckage flowed downstream with the current, and men clung to anything that would float. Pitiful shrieks, cries, and entreaties for help pierced the night. One survivor later would recall that the horrible scene reminded him of stories of doomsday, the end of the world.

The rising sun on 27 April illuminated a terrible spectacle. Some survivors were holding on to logs, barrels, pieces of decking, and anything else that would float. Swept downstream by the current, some had floated all the way to Memphis. Boats of all sizes scoured the river and pulled live men out of the cold waters. Several hundred lives thus were saved, but many rescued men were badly burned and injured. One former Confederate soldier, alone in a small boat, made trip after trip and plucked 14 debilitated men from the river.

Rescuers took about 600 survivors to doctors and to the hospital in Memphis. Of those, over 200 died within a few days, mostly as the result of burns, other bodily trauma, and general debility. For a week a river barge left Memphis each morning to search for bodies. Each evening it returned with its gruesome cargo. For over a month at locations downstream, corpses would wash up onto the riverbanks.

Cause of the Tragedy: *Sultana* had been heavily overloaded, had defective boiler repairs, and had been fighting a strong current. Three of the four boilers had exploded. The steamboat burned and sank, and about 1,800 people perished. During an investigation the cause of the tragedy was determined to be fourfold:

– Mismanagement of water levels in the four boilers.
– A severely overloaded vessel.
– A top-heavy vessel.
– Faulty boiler repairs a few days earlier.

As the steamboat had clawed its way upriver it naturally had to follow twists and bends in the channel. As the top-heavy vessel turned, it listed heavily away from the direction of each turn. The four interconnected boilers had been mounted side-by-side. As the steamboat listed during a turn, water naturally had drained out of the higher boilers and flowed into the lower ones.

During a turn with coal fires in fireboxes still heating the partially empty boilers, extreme "hot spots" were created. Then, when the boat turned in the opposite direction, water would rush back into the partially empty boilers. When this water hit "hot spots" it would instantly flash into steam. This created sudden increases in steam pressure that eventually exceeded the boilers' strength, resulting in an explosion. The combined effect of side-to-side listing, low water level in the boilers, and faulty repair work proved to be more than the boilers could withstand.

Sabotage?: Notwithstanding the officially stated cause, conspiracy theories quickly flourished. Perhaps the most credible is one that did not surface until 23 years after *Sultana* burned and sank. In 1888 a St. Louis resident alleged that his former business partner, Robert

Louden, had made a deathbed confession. Louden supposedly told his partner that he sabotaged the steamship with a coal torpedo.

During the war Louden had been a known Confederate agent and saboteur. With regard to the *Sultana* he had a motive, he had the opportunity, and he had the expertise to destroy the steamboat. During the war, coal torpedoes had been developed by Capt. Thomas Courtenay of the Confederate Secret Service. Beginning in January 1864 they had been manufactured at the Seventh Avenue Artillery Foundry in Richmond.

A coal torpedo (years later it would be called a *mine* or an *IED*) looked like a lump of coal. In fact it was an irregular-shaped hollow iron container filled with six ounces of gunpowder. After it was filled with gunpowder it was closed with a threaded iron plug, dipped in beeswax, and then rolled in coal dust. The finished coal torpedo looked exactly like an innocent lump of coal.

In use as a sabotage tool, a coal torpedo would be surreptitiously added to a ship's coal supply. When the vessel was underway the explosive coal torpedo, along with genuine coal, would be shoveled into a firebox under one of the boilers. Burning coal would heat the coal torpedo and cause the gunpowder to explode. This explosion would rupture the boiler above, causing catastrophic damage. This usually would sink the vessel or cause it to be destroyed by fire.

During the war Courtenay had been tasked with putting coal torpedoes in coal stockpiles used by Union ships. After the war he would travel to England and try to sell his "secret" to the British. However, by then the sabotage device was no longer a secret, and Courtenay never made a sale.

Louden's alleged verbal claim remains controversial. He indeed had been involved in coal torpedo plots against Union river shipping during the war. However, most technical experts dismiss his alleged deathbed statement as simple bravado.

Search for a Scapegoat: Why had the steamboat been so packed with humanity, so overloaded? Many people blamed LtCol. Hatch, but he had political connections. He often displayed letters of commendation he had received from Abraham Lincoln and Gen. Ulysses Grant. In any event, shortly after the *Sultana* tragedy he had left the army and was no longer accountable to military authorities.

Further, no civilian judge wanted to put him on trial.

The U.S. Army naturally was loath to publicize the disaster and tremendous loss of life. There was an official inquiry, but nobody paid much attention to it. A U.S. Army report dryly noted:

> The great power of steam, having once found a yielding place, tears everything before it, producing an explosion. . . . The explosion of a boiler is caused by water coming in contact with red-hot iron . . . and the destruction of boilers and the boat that carries them is the consequence.

One U.S. Army officer, Capt. Frederick Speed, initially was charged with overloading the vessel. However, all charges were dismissed when an army judge advocate general learned that Speed had been in the parole camp all day. He had nothing to do with loading the steamboat. Capt. Mason and his officers all perished on the vessel, so they could not be held accountable. In the end there was no civil or military trial, and the tragedy on the Mississippi River faded away into the foggy mist of history.

Death Toll: Loss of the steamboat *Sultana* got limited newspaper coverage, for it was totally overshadowed by other events. The long bloody war was ending. Lincoln had been assassinated and had been dead for less than two weeks. His assassin, John Wilkes Booth, had been cornered and shot the previous day. Also on the previous day the last major rebel force, commanded by Gen. Joseph Johnston, had surrendered. All major newspapers concentrated on such events, and little attention was directed to the maritime disaster in the west.

The exact number of deaths remains unknown, for a passenger manifest was not created prior to leaving Vicksburg. The official tally of 1,800 deaths was developed by the U.S. Customs Service. It is an educated estimate. There were about 450 known survivors. Together those two totals mean that about 2,250 souls were on the vessel when it left the Memphis dock. Many of the dead are buried in Memphis National Cemetery. Families claimed some others and carried them to hometowns and other locations for burial.

Although the exact number of persons who died is not known, the toll is certainly greater than the 1,512 deaths attributed to the *RMS Titanic* steamship disaster 47 years later. In terms of number of lives lost, the *Sultana* tragedy still remains the worst maritime disaster in United States history.

Survivors: In 1885, twenty years after *Sultana* sank, survivors from Indiana, Michigan, and Ohio formed the National Sultana Survivors Association. They began meeting annually as close to the 27 April date as practical. By 1925 advancing years had taken their expected toll, and only a dozen elderly men were still able to make the trip. Four years later in 1929 only two men showed up. The last known survivor, Pvt. Albert Norris, age 94, died at home seven years later on 9 January 1936. That was 71 years after the steamboat took so many unsuspecting young soldiers to fiery and watery graves.

> They wrote in the old days that it is sweet and fitting to die for one's country. But in modern war, there is nothing sweet nor fitting in your dying. You will die like a dog for no good reason.
> *[Ernest Hemingway, Notes on the Next War, 1935]*

Sultana at the dock, loaded with former Union POWs.

River steamboats smaller than Sultana loading and unloading cargo at a landing in Memphis, Tennessee, circa 1870.

Fanciful sketch of the fiery destruction of Sultana (from Harpers Weekly, 20 May 1865).

Photograph of a "coal torpedo" without its threaded plug.

The Soldier Atop the Memorial

Benson was a legend for his Civil War exploits and his postwar dedication to worthy causes.
[Edward J. Cashin, introduction to Berry Benson's Civil War Book, 1991]

Overview: During the American Civil War hundreds-of-thousands of combatants were captured by the enemy. Relatively few of them escaped from prisoner of war camps. Many more thousands were severely wounded or maimed. Those sent home to recover from such debilitating injuries often failed to return to their military unit.

After the war ended most survivors led productive but quiet lives. There were exceptions, of course, and one stands head and shoulders above the masses. He was a most unique Confederate soldier, and after the war he became an extraordinary citizen.

Early Life: Berry Greenwood Benson, eldest son of Abraham and Nancy Benson, came into the world on 9 February 1843. He spent his formative years in Hamburg (later, *North Augusta*), South Carolina, which had been founded in 1822 by a German immigrant. In 1849 when Benson was six years of age his father spent two years searching for a fortune in California gold fields. Not finding one, he returned to Hamburg and worked as an accountant.

Publicly funded education was not available. However, Benson's father paid tuition for his son to attend the David Griffin school in Augusta, Georgia, across the Savannah River from Hamburg. Berry walked three miles to school each day, crossing the river on a railroad trestle. From an academic standpoint the Griffin school far surpassed other colleges and academies of the day. Benson learned to read Latin, Greek, and French. He studied history, literature, natural sciences, and geography, but advanced mathematics became his favorite pursuit. Further, he was an accomplished speller, a trait that would serve him well in adulthood. When not attending classes he evolved into an avid fisherman and hunter.

When Benson was 17 years of age his state seceded from the United States on 20 December 1860. Swept up in the martial spirit of the times, the teenager joined the local militia unit, the Hamburg Minutemen. His brother, Blackwell Benson, was only 15 at the time, but he joined with his older brother. The Hamburg Minutemen soon became part of the 1st South Carolina Volunteer Infantry Regiment. The Benson brothers went with their regiment to Charleston and helped man the Edgefield Battery, a coastal artillery unit. In April 1861 they took part in the bombardment of Fort Sumter.

Life in the Confederate Army: After a quick Union surrender at Fort Sumter the Benson brothers entrained for Suffolk, Virginia, to join C.S. Army forces there. Berry Benson's six-month enlistment expired, so he reenlisted for the duration of the war and was promoted to the rank of corporal. During the next year, 1862, he fought during the Seven Days Battles, at Second Bull Run (aka, *Second Manassas*), Sharpsburg (aka, *Antietam*), and Fredericksburg.

> The air was horribly filled with the stench of decaying bodies. . . . If a man wants to see Hell upon earth, let him come and look into this black, bloody hole – upon this horrid confusion, this reeking mass of corruption of rotting corpses.
> *[Berry Benson, diary entry, Spotsylvania aftermath]*

During the winter of 1862-1863, Berry Benson and his brother honed their reputation as skilled hunters and marksmen. Because of their demonstrated ability they were reassigned to a new unit, an elite sharpshooter battalion. The C.S. Army issued each battalion member an ultra-accurate imported .577 caliber British Enfield rifle.

In the spring of 1863 as part of the Army of Northern Virginia, Benson and his sharpshooter comrades prepared to defend Virginia from a Union invasion at Chancellorsville. On 2 May as part of Gen. Thomas "Stonewall" Jackson's corps, they surreptitiously marched 12 miles around and behind the Union Army right flank. Late in the afternoon they charged unsuspecting Union troops. Even though they were outnumbered roughly three to one, southerners routed Union forces. By noon the next day the Confederate victory was complete. Later termed the "perfect battle," success at Chancellorsville became the pinnacle of Confederate military prowess.

Benson Is Wounded: Berry Benson paid a heavy price. During the Confederate charge a Union Minie-ball had crippled him when it tore through his left leg below the knee.

> I had fired some half dozen shots when I felt a blow on my left leg. Stooping down to see, I found my stocking filling with blood.
> *[Berry Benson, diary entry for 2 May 1863]*

Surgeons did not amputate. After Benson was stabilized in a field hospital he traveled by train to Richmond and arrived on 7 May. After a brief stint in a small hospital he was admitted to the huge Howard's Grove military hospital. Due to the severity of his wound, on 1 June he was transferred to a military hospital near his home. He reached Augusta, Georgia, via train and was admitted to the hospital that had been established in the former Eagle & Phoenix Hotel on Broad Street. There he slowly recuperated and regained his strength. Four months later he would write in his diary: "I still lament that I was not at Gettysburg *[in July 1863]*."

> Wounded at Chancellorsville, Benson missed the Battle of Gettysburg but *[would be]* back in the ranks by winter.
> *[Joseph G. Bilby, in American Rifleman, 2017]*

The Augusta to which Benson returned had changed from his schoolboy days. In Augusta the huge Confederate Powder Works supplied almost half of all gunpowder used by the C.S. Army. Injured, sick, and wounded soldiers were housed in churches, hotels, and the military hospital. Benson hobbled around Augusta using a crutch, and he helped his father with the accounting business.

Many wounded soldiers in Benson's situation would have been tempted to stay at home longer, but Benson was different. By autumn he thought he had recovered enough to rejoin the army. He caught a train headed for Fredericksburg, Virginia, where his comrades were encamped for the winter. There he learned that he had been promoted to the rank of sergeant. In camp during that winter of 1863-1864, Benson completed his medical recovery. He had time on his hands, so he read *Les Miserables*, *Paradise Lost*, *Tannhauser*, *The Life of Mahomet*, and other scholarly works.

In the spring of 1864, Benson became one of the battalion scouts.

He fought in the bitter Battle of the Wilderness and at Spotsylvania. Between battles and campaigns he often served as a picket, and he wrote in his diary:

> We became quite friendly with the enemy's pickets posted on the opposite side *[of the river]* and used to talk with them and exchange newspapers. . . . Once a Federal band came down to the river and played "Dixie." We cheered them vociferously, of course. Then it played "Yankee Doodle" and the enemy cheered. Then *[it played]* "Home, Sweet Home" and the cheer went up loud and long from both sides of the river.

First Capture and Escape: While he was scouting inside Union lines near Spotsylvania on 16 May 1864, enemy soldiers cornered Benson and took him prisoner. Seven days later on 23 May, Union guards took him to Point Lookout Prison Camp in Maryland.

The prison camp had been established on a sandy peninsula on the Chesapeake Bay shore. Benson noticed that latrines had not been dug in the sandy soil. Instead, privies had been built in the bay near the shore in water about chest-deep. Wooden piles supported each privy, and each had wooden sides about four feet high in the interest of modesty. Wooden walkways connected each privy to the shore. In the daytime, prisoners could use the walkway and enter one of these privies whenever they wished. An escape opportunity!

Benson would remain a prisoner for only two more days. On 25 May as twilight began to settle over the prison camp he sauntered into one of the privies and crouched down, out of sight of the guards on the shore. After 30 minutes he correctly guessed that the bored guards had forgotten him. In the darkness he slipped over the low wall behind the privy and lowered himself into the Chesapeake Bay.

> *[In the dark]* I went into deeper and deeper water. When it came up to my chin I turned and began to wade up the bay, *[and]* a moment later I was swimming.
> *[Berry Benson, diary entry for 25 May 1864]*

Benson waded and swam for two miles, and then he walked ashore onto Maryland soil far north of the prison camp. After making his way northward he came to a place where the Potomac River, southwest of him, was more narrow. He swam across the river to Virginia and began blindly walking toward friendly lines.

Second Capture and Escape: Benson had no map, so he blundered southward as well as he could. As he passed farms and gardens with row crops he "borrowed" vegetables to fill his stomach. Chicken pens were a good source of eggs, which he ate raw to retain his strength. He took live chickens also, and at night in the woods he cooked them over a small fire.

> Passing thro' a cornfield, I put some of the ears in my pockets. Then I came to an orchard, and oh, how delicious the apples were!
> *[Berry Benson, diary entry, post-escape foraging]*

Meandering southward in the dark at night, Benson had the misfortune to stumble into Union pickets. On 31 May 1864 he once again was captured. He admitted to his captors that he had escaped from the prison at Point Lookout, so they decided not to take further chances with him. They took him directly to Old Capitol Prison in Washington, DC. An escape from there would be highly improbable. However, on 23 July two months later, Benson chanced to be among 662 Confederate POWs who were put on a train bound for the Union prison camp at Elmira, New York.

Elmira Prison Camp was new. The former Union training base had been converted into a prison. The first 4,000 prisoners had been housed in former Union barracks. Additional arrivals were housed in little A-tents, three men to a tent. The camp at Elmira had become grossly overcrowded, and there were not enough tents to go around. With winter only three months away, many prisoners had to sleep on the ground with no protection from the elements. Such privations caused prisoners to call the camp "Hellmira." Benson already had escaped once, so he resolved to try again.

Benson and nine other prisoners (John Maull, Washington Traweek, John Putegnat, Gilmer Jackson, William Templin, S.C. Malone, Glenn Shelton, James Crawford, and J.P. Scruggs) began digging a tunnel, but it was discovered by Union guards. They elected to try again. Their new tunnel began inside an A-tent, and when complete it would extend 67 feet to the outside of the stockade wall. Benson and his compatriots used metal saucers and case knives to dig each night. He used his mathematical skills to guide the tunneling endeavor. They put excavated dirt into their haversacks and secretly spread it around the camp during the daytime. Under cover of darkness the ten men crawled through their tunnel to freedom on 6 October 1864.

Back to the War: Some of the escapees wanted nothing more to do with warfare, so they headed for the nearby Canadian border. Berry Benson was different. He walked southward, hopping rides on boxcars when he could. Several times he passed himself off as a Pennsylvania farmer. Once he feigned sorrow and told an elderly passenger train conductor a concocted tale about trying to get home to see his dying sister. The sympathetic conductor let Benson ride free of charge in a passenger coach. After three weeks he reached Confederate lines at the end of October.

> Benson had walked south, determined to rejoin his unit. Incredibly, he reached the Confederate lines in the Shenandoah Valley in late 1864.
> *[Joseph G. Bilby, in American Rifleman, 2017]*

The C.S. Army rewarded Benson with a furlough, so he took a train southward to Hamburg to visit his family. While there he learned that a Union army commanded by Gen. William Sherman had left Atlanta and was driving southeast across Georgia. No one knew where Sherman was headed, but the City of Augusta, across the Savannah River, seemed to be in his crosshairs. Benson joined Confederate troops entrenched to defend Augusta. However, by Christmas in 1864 it became obvious that Sherman had entered the City of Savannah on the coast. He had bypassed Augusta.

Benson was eager to get back in the war. He caught a northbound train and rejoined his sharpshooter comrades at Petersburg, where he arrived on 25 January 1865. That winter and in the early spring of 1865 bitter battles raged along the massive Confederate defenses protecting Petersburg and Richmond. After Confederates repulsed one Union attack Benson wrote in his diary:

> *[I walked among]* dead bodies . . . swollen to twice the size of *[living]* men. The air was horribly foul with the

> stench of decaying bodies. In some places they lay so thick on the ground that I had to pick my way amongst them and shun the fearful gaze of their open eyes. I held my head up, looking ahead, afraid to look down.

When Richmond was evacuated by the Confederate Army and Confederate government, Benson stayed with Gen. Lee's army. At Appomattox when told that Lee had surrendered, Berry and his brother vowed to continue the fight. The Benson brothers did not surrender or sign the Amnesty Oath. That night they slipped out of camp and made their way south to North Carolina where they planned to join the army of Gen. Joseph Johnston. However, upon arrival they found that Johnston also planned to surrender. The Benson brothers left North Carolina and walked all the way to Hamburg. The war was over. Yet, they still carried their British Enfield rifles, and they never surrendered.

> Not a drum was heard anywhere. Never a bugle blew when we came home from the war.
> *[Berry Benson, post-war diary entry]*

Early Adulthood in Augusta: After returning to Hamburg, Benson first worked as a schoolteacher in the nearby Sweetwater Church School. He soon became a cotton merchant, classifying fibers in each bale, buying the bales, and selling them to cotton mills. Following in his father's footsteps and in his own interest in mathematics, he began working as an accountant. All textile mills in Augusta and Hamburg hired him to do their auditing. He began writing scholarly articles about the war, and the *Augusta Chronicle* newspaper published them. He also wrote philosophical articles that were published by *Outlook Magazine* and *Century Magazine.*

Benson developed a unique bookkeeping and accounting system. He named it the Zero System and marketed it nationally. In 1868 he married Jeanie Oliver, and together they would have six children, two boys and four girls. Berry and Jeanie wrote and published poetry, and their four daughters would become accomplished pianists. One studied violin in New York and became a concert performer.

During the Reconstruction Era, violence in the Augusta area had become common. On 8 July 1876 hundreds of armed vigilantes stormed into Hamburg and shot six state militiamen in what would become known as the Hamburg Riot. Vigilantes captured 25 more militiamen who had been hiding in crawl spaces under houses in the town. Berry Benson had gained a reputation as a man who could resolve disputes. He quickly was summoned, and he arbitrated the release of captive militiamen.

> I have never known a finer man, or a better man, than Berry Benson.
> *[excerpt from an editorial in the Augusta Chronicle on 3 January 1923]*

Benson and the Confederate Memorial: Benson's civic work and his military service during the war, plus his newspaper articles, kept him in the public spotlight. He became well known as the soldier who had been captured twice, imprisoned twice, and escaped twice. Plus, he always had returned to the fray, and he never surrendered. His name came to mind when plans were begun for an enormous Confederate Memorial in Augusta.

In 1875, ten years after the war had ended, the Ladies Memorial Association began raising money to fund a massive war memorial. The structure would honor soldiers who died during the war. After three years of planning and fund-raising, construction began in the median of the 700 block of Broad Street in Augusta.

The memorial would cost $17,331.35, a princely amount in the 1870s. The structure would tower 76 feet up into the sky. The Philadelphia architectural firm of VanGruder & Young would design it. The massive Georgia granite base would be topped by a shaft of Carrara marble. Famed Italian sculptor Antonio Fontana would carve the stone. Surrounding the base would be life-size statues of four Confederate generals: Robert E. Lee, William Walker, Thomas "Stonewall" Jackson, and Thomas Cobb. Atop the marble shaft would stand a life-size statue of an anonymous common soldier. Berry Benson was chosen as the model for that soldier.

Amid much pomp and ceremony the Confederate Memorial was dedicated on Halloween, 31 October 1878. It was the largest such memorial in the state of Georgia. The *Augusta Chronicle* reported that over 10,000 citizens attended the ceremony. High atop the marble shaft the anonymous soldier, modeled after the likeness of

Sgt. Berry Benson, gazes southward along Broad Street.

Civic Involvement: While Benson was working as the auditor for textile mills he rubbed shoulders with mill management each day. Yet, he sympathized with rank and file mill workers during the textile workers strike in 1898. During the strike he convinced mill managers in Atlanta to hire destitute striking workers from Augusta. He authored newspaper articles lamenting the plight of mill workers and became the chief arbitrator responsible for ending the strike.

> Benson's post-war career was compelling like his military service. . . . While working as an auditor in a textile mill he evolved into a vocal advocate for striking mill employees. *[Joseph G. Bilby, in American Rifleman, 2017]*

Always fascinated by mathematics, Benson cracked the secret French Army military code and shared his findings with the U.S. War Department. During the Spanish-American War he volunteered to work for U.S. Army code-breakers. Always fascinated by spelling, he became a proofreader for *Webster's Unabridged Dictionary* and *Funk & Wagnall's Dictionary*. He also helped a former Union Army soldier gather information for a book about Elmira Prison Camp.

Benson and the famed Leo Frank Case: Benson garnered national recognition for his involvement in a famous anti-Semitism judicial case. A Jewish plant manager, Leo Frank, at the National Pencil Company in Atlanta was charged with the April 1913 murder of a 13-year-old female employee, Mary Fhagan. In an overly obvious rush to judgement, Frank was convicted and sentenced to death.

Intrigued with the case, Benson studied courtroom transcripts and discovered multiple flaws in testimony of the crucial witness, an ex-convict. Further, Benson discovered that the witness had provided four contradictory affidavits to prosecutors. Newspapers throughout the United States published Benson's research. Benson sent his findings to John Slaton, governor of Georgia, and the governor concluded that Frank indeed was an innocent man.

> *[Slaton]* commuted the sentence to life imprisonment, assuming that Frank's innocence would eventually be fully

established and he would be set free.
[Leonard Dinnerstein, in the New Georgia Encyclopedia, 2003]

Tragically, before the legal process ran its course, vigilantes struck. On 16 August 1915 they lynched Frank. Over 70 years later in 1986 the Georgia State Board of Pardons and Paroles would posthumously pardon Frank. In part, the board's rationale was "the State's failure to protect the person of Leo M. Frank and preserve his opportunity for continued legal appeal."

Benson's Latter Years: During World War I, Benson and his wife sympathized with the plight of thousands of French war orphans. They "adopted" five of them by paying for their care and convinced residents of Augusta to adopt 160 others. At age 74, still physically fit and mentally alert, Benson led the Georgia Battalion during the 1917 Confederate Veterans parade in Washington, DC.

Always interested in finance and mathematics, Benson discovered the scheme of Carlo Ponzi, an Italian swindler and con man. Ponzi had begun buying discounted postal coupons in other countries and redeeming them at face value in the United States. He promised investors a 100 percent profit within 90 days. However, the profit was illusionary. Ponzi actually paid early investors not with profits, but instead with money from later investors. Benson researched and understood the ruse and alerted the Massachusetts Attorney General. Newspapers credited Benson with initiating the investigation that led to Ponzi's conviction and imprisonment (the illicit tactic later would become universally known as a "Ponzi Scheme").

Benson remained active, and he led the local Boy Scout troop on 15 mile hikes. He evolved into a tireless community leader. With regard to his military service, he is considered to be the most famous survivor of Elmira Prison Camp. He is the loyal Confederate soldier who was captured twice and escaped twice, the dedicated soldier who never surrendered.

Benson spent his latter years with his family in his hometown, which has been renamed North Augusta. When he was 79 years of age he died at home on New Years Day in 1923. His mortal remains are buried in Sunset Hill Cemetery in North Augusta. In accordance with his wishes the tombstone inscription is simple:

BERRY BENSON
Sergeant 1st S.C.
Inf*[antry]* C.S.A.

The former Susan Williams, daughter of Leon Judson Williams, had been born in rural Plum Branch, South Carolina, in 1892. In adulthood she married Berry Benson's son, Charles Benson. As Berry Benson's daughter-in-law, Susan Williams Benson resolved that Sgt. Benson's military and civic service would be remembered. She typed the portions of his diary that related to his military service. Her project was first published in 1962 as *Berry Benson's Civil War Book*. For the later printings a subtitle was added: *Memoirs of a Confederate Scout and Sharpshooter.*

Berry Benson has not been forgotten. The Sgt. Berry Benson Camp, Sons of Confederate Veterans, is named in his honor. The *Augusta Chronicle* often references both his wartime service and his many contributions to society. Most recently, in 2014 the newspaper reported that a huge new flagpole has been dedicated adjacent to his grave. And of course, in the Augusta area Sgt. Berry Benson will be forever remembered as "the man atop the monument."

> Who knows? But it may be given to us after this life to meet again in the old quarters to play chess and draughts, to get up soon to answer the morning roll call, to fall in at the tap of the drum for drill and dress parade, and again to hastily don our war gear while the monotonous patter of the long *[drum]* roll summons us to battle.
>
> Who knows? But again the old flags, ragged and torn, snapping in the wind, may face each other and flutter, pursuing and pursued, while cries of victory fill a summer day? And after the battle, then the slain and the wounded will arise, and all will meet together under the two flags, all sound and well, and there will be talking and laughter and cheers, and all will say: "Did it not seem real? Was it not as in the old days?"
>
> *[Sgt. Berry Benson, writing circa 1880 and quoted over a century later by Shelby Foote in the television series, The Civil War.]*

Grave marker for Sgt. Berry Benson in Sunset Hill Cemetery, North Augusta, South Carolina (photo by the author).

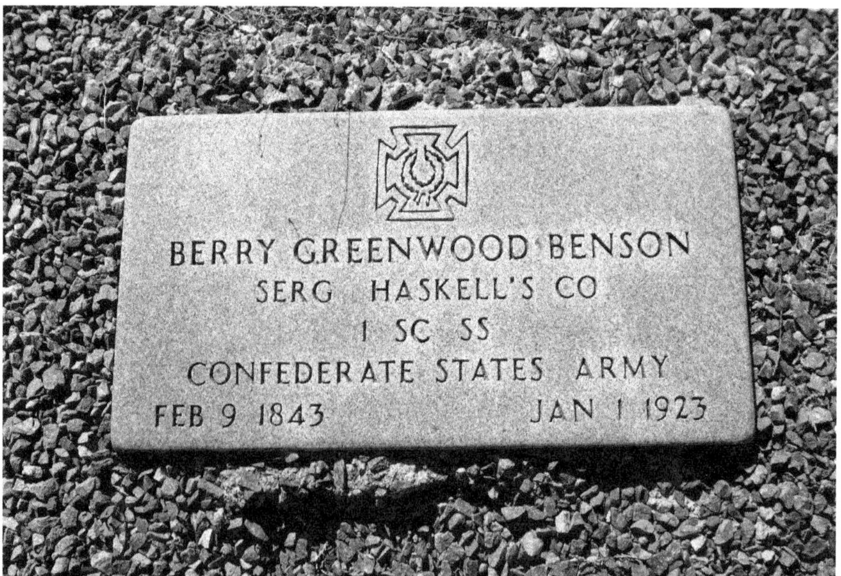

Additional grave marker, provided by the U.S. Veterans Administration, for the grave of Sgt. Berry Benson (photo by the author).

Confederate Memorial on Broad Street median, Augusta, Georgia. It was dedicated on 31 October 1878 (photo by the author).

An anonymous Confederate solider stands atop the memorial in Augusta, Georgia. The model for this soldier was Sgt. Berry Benson (photo by the author).

The Tragedy of the Reconstruction Era

> Physical separation of the races was the most profound and unfortunate consequence of the Reconstruction Era. Caucasians and Negroes separated and built their communities, schools, churches, and cultures largely apart from each other. This separation, instigated by the United States Government, would last for well over a century.
> *[Walter Edgar, South Carolina, a History, 1998]*

A Devastated Society: In states of the former Confederacy the long Reconstruction Era (1866-1877) played out against a backdrop of ruination and devastation. Small farms had fallen into disuse, and emancipation of slaves had doomed tobacco and cotton plantations. Most horses, mules, cattle, and hogs had been stolen or eaten during the war. Railroad trestles, tracks, repair yards, and rolling stock had been worn out or destroyed by the U.S. Army. The former coin of the realm, the Confederate dollar, was worthless. Former investments in slaves also were worthless. Most citizens were forced to substitute bartering in lieu of monetary transactions.

Most former slaves still lived on plantation tracts or medium size farms where they had lived and worked in the past. No longer treated as refugees or fed by the U.S. Army, they discovered that food was scarce or non-existent. Many of them ventured into cities in search of employment, food, and shelter. However, jobs were scarce. Other than food, there was little with which to repay laborers. Former slave women fared somewhat better than men in the job market. Many found work as live-in cooks, maids, child nurses, and laundresses, but food and board usually were the only forms of payment.

Much of the South had degenerated into a poverty-stricken backwater. The obvious solution lay in restoration of farming. Yet, landowners who had survived the war had to mortgage their land to get funds needed to replant and begin farming again. There was no money to pay laborers, so most landowners resorted to sharecropping, often called tenant-farming.

Sharecropping details varied, but in general a landowner would authorize a destitute laborer to farm a given tract of land. The landowner often furnished seed, fertilizer, and occasionally use of a mule for plowing the fields. Sometimes the sharecropper also got a cabin in which to live. When crops were harvested and sold the landowner and sharecropper would share the profits. This system proved to be satisfactory for both parties. Often it was the only way they could put food on their tables.

Start of the Reconstruction Era: Before his assassination the U.S. President, Abraham Lincoln, had planned to expedite readmission of southern states into the Union after the war ended. Until his death the moderates in Congress thought the process would be simple. They had assumed that each rebel state would (1) repeal its ordinance of secession, (2) ratify the Thirteenth Amendment abolishing slavery, and (3) grant full civil rights to all adult men. Presumably these former rebel states then would resume their pre-war roles in the U.S. Congress and in the union of states.

Unforseen events would disrupt Lincoln's peaceful scenario. The next 12 years would become known as the Reconstruction Era, a tragic period in United States history.

> There is only one sure and safe policy for the immediate future. The North must remain the dictator of the republic until the spirit of the North shall become the spirit of the whole country. The South's treason is still unpunished.
> *[editorial in a Radical Republican newspaper]*

After Lincoln's assassination the mood in Congress changed. Bent on revenge for the president's death and the human cost of the war, Congress decided to punish former rebel states. Midterm national elections in 1866 gave Republicans a majority in both the Senate and House of Representatives and decisively changed the balance of power. With over two-thirds majorities in both houses of Congress, Republicans could override any presidential veto.

So-called Radical Republicans controlled Congress and passed new legislation entitled, An Act to Provide for the More Efficient Government of the Rebel States. Often called the Reconstruction Act, it applied to ten southern states to which U.S. Army soldiers would be deployed. Congress voided existing state governments in these states and placed them under U.S. Army control. Congress also set up five military districts in the South:

District 1 – Virginia
District 2 – South Carolina and North Carolina
District 3 – Georgia, Florida, and Alabama
District 4 – Arkansas and Mississippi
District 5 – Texas and Louisiana

The U.S. Army registered eligible male voters (women of that era could not vote) and supervised new elections. The result was totally predictable. Randolph Campbell, an historian, would write:

> The Reconstruction Act called for registering all adult males to vote, Caucasian and Negro, except those who had engaged in the rebellion. This barred from registration most men who had served in the Confederate Army, all county and city officials and officeholders, and even minor functionaries such as sextons of cemeteries.

The United States deployed slightly over 21,000 army soldiers to enforce the Reconstruction Act. Four "border states" and "slave states" (Kentucky, Missouri, Maryland, and Delaware) had not joined the former Confederacy and were not subject to martial law during Reconstruction. Tennessee, which already had been readmitted to the Union, also was not subject to Reconstruction. Neither was the new state of West Virginia, which had seceded from Virginia in 1863 and had joined the Union.

Carpetbaggers and Scalawags: Congress authorized a Freedmen's Bureau to assist former slaves in adjusting to life in a new free labor economy. Yet, few former slaves initially participated at the ballot box. Like many poor Caucasian farmers and laborers, most former slaves were illiterate. Seeing this golden opportunity, educated men from the North, called *carpetbaggers*, rushed to southern states to run for public office. They staffed the new Freedmen's Bureau and encouraged former slaves to participate in the political process. In the South men and women of both races who supported these Radical Republicans were termed *scalawags*.

Elections supervised by the U.S. Army resulted in a Radical Republican regime composed of carpetbaggers, scalawags, freedmen, and their supporters. Together they formed an ironclad voting block.

These radicals took over virtually all state and local governments and held total political power in the South. They drafted new state constitutions and controlled the judicial system, the banking industry, and government at all levels.

The U.S. Congress would adopt two constitutional amendments during the Reconstruction Era. The Thirteenth Amendment, ratified in December 1865, already had abolished chattel slavery throughout the United States. The Fourteenth Amendment, ratified in 1868, granted citizenship to persons born in the United States. The Fifteenth Amendment guaranteed the right to vote to all adult men regardless of their prior servitude status.

The Radicals were realists. They knew their glory days would not last forever, for sooner or later those who had supported the former Confederacy would regain the right to vote. When that happened, the Radicals knew, they would be voted out of office. Therefore most of them viewed their tenure in office as a transient opportunity to enrich themselves at public expense.

Radical southern state legislators lavished perks upon themselves. In Columbia, South Carolina, the Speaker gave himself $1,000 bonuses and spent $1,600 on brass spittoons. Some tax money was earmarked for county projects, but none actually was spent to benefit the public. In one rural area there was no way for wagons and buggies to get across a large creek to reach the county seat. All bridges had been washed away, and neither the state nor the county would replace them. One resident complained in a letter that was published in the *Edgefield Advertiser* newspaper:

> Sixteen hundred dollars for spittoons for the legislature, and one thousand dollars extra to Speaker Moses, but not a bridge across Stevens Creek for the hard working people of Edgefield County – not a single bridge across Stevens Creek from its mouth to Dorn's Mines, a distance of forty miles. Conceive of it! Conceive of the inconveniences and annoyances suffered by our fellow citizens . . . .
>
> And with a tax of three mils for county purposes! Where are the County Commissioners? Have they no idea of the requirements of their position? No sense of duty? No energy? No conscience? No mercy?

Property taxes skyrocketed up to ten times the pre-war rate. Cash-poor farmers rarely saw 20 dollars in U.S. currency during an entire year. Nonetheless they had to pay their property taxes or lose their land. Graft and corruption ran rampant in state governments. Millions of tax dollars went to school commissioners who funneled much of it into their own pockets, while kicking back a healthy percentage to legislators who made it all possible. A large percentage of tax money funded railroad subsidies. Such subsidies, often given as bonds, went to industrialists, speculators, and financiers in the North. In return the grantors got rewarded with extravagant fees which enabled them to purchase bonds for themselves and take their families on lavish excursions to Europe.

Fleeing from the Reconstruction Era: Some able-bodied veterans and their families saw the proverbial handwriting on the wall. Faced with military occupation of the South by the U.S. Army, they fled. Some went to England and Canada, both of which are English-speaking countries. Others sought refuge in California and other western regions that would not be under the heel of occupying U.S. Army troops.

Other ex-Confederates fled southward to the colony of British Honduras (later, *Belize*) in Latin America. British citizens there had been sympathetic to the Confederate cause during the war and had sold munitions to the South. After the war the British colonial governor recruited former Confederates who were experienced in growing cotton and sugarcane. He granted tax breaks and subsidies to those willing to settle in the English-speaking colony.

Other Confederates went to Mexico in 1865 and founded Carlota, between Mexico City and Veracruz. Matthew Maury, an oceanographer and former Confederate soldier, was a friend of Emperor Maximilian. The emperor and Maury planned not only Carlota, but also agricultural colonies near Monterrey and Tampico. The emperor offered land to those willing to settle and become Mexican citizens. Unfortunately, two years later the emperor was captured and executed by rebels. Thereafter no more land was granted, and Confederate colonists already in Mexico blended into the local population.

The most popular destination for fleeing former Confederates proved to be Brazil. Its government sought educated immigrants and offered subsidized travel to Brazil, tax breaks, and cheap land. The first disillusioned Confederate to book steamship passage to Brazil was William Norris of Alabama. Accompanied by 30 families, he docked in Rio de Janeiro in December 1865. He and his entourage settled in the Sao Paulo region and founded what natives called Villa dos Americanos (meaning, *Village of the Americans*).

The new Brazilian colony introduced modern farming techniques for cotton and food crops. As several thousand more ex-Confederates joined them, they established the first Baptist church in Brazil and built a school for their children. Called *Confederados* by Brazilians, the colonists prospered. Years later in 1924 their growing city would become a municipality and be renamed Americana.

Over passing years at least 3,585 ex-Confederates would move to Brazil, and most would make Americana their home. Although the expatriates soon would speak Portugese and intermarry with native Brazilians, the Confederate influence would remain strong. The annual Festa Confederada would draw thousands of revelers, and the American Descendants Association would promote their American heritage. A large Confederate monument bearing names of original ex-Confederate founders would stand near the city center.

Schools – A Double-Edged Sword: In the South amid the fraud and corruption there were glimmers of hope. Before the war almost no public schools had been funded by state governments. Parents who could afford the cost had sent their children to parentally funded schools. Some slave children had learned the basics of reading and writing in "schools" taught by their Caucasian playmates, but formal schools for slaves had been prohibited by state laws. However, in the post-war era Radical Republican legislators made a commitment to government schooling for all children, supposedly a noble goal.

Some new schools were built in cities, but Radical legislatures mandated that these schools be segregated by race, exclusive of schools in and around New Orleans. Further, it was assumed that in rural areas, money spent on schooling would be wasted on children destined to become farmers or common laborers. Teachers in urban schools usually were poorly qualified and poorly paid. In schools, children would get academic basics plus "cultural training" based upon the Radical Republican agenda. Commissioners hoped that children would pass these concepts along to their parents, who then would vote to maintain the status quo.

Both races had attended the same churches together before the Reconstruction Era had begun. Now the racially segregated school concept began to spill over into southern churches. Prompted by the Freedmen's Bureau, Negro church members began leaving to form their own churches. The two primary denominations, Methodists and Baptists, tried to stem the massive Negro flight from old established churches. Their efforts failed. Within a decade southern churches, exclusive of some Catholic churches in Louisiana, would become totally segregated by race. Societal unity had been sacrificed on the altar of Reconstruction.

Democratic Clubs: The former ruling class in the South had lost its voice in the political process. Taxes rose, and the price of cotton declined. Government corruption became so commonplace that officeholders no longer made attempts to conceal it. Gradually responsible citizens concluded that common decency and order could be restored only by force. Declaratory words of Thomas Jefferson began appearing in newspaper notices and on public handbills:

> Governments are instituted among men, deriving their just powers from the consent of the governed. Whenever any form of government becomes destructive of these ends, it is the right of the people to alter or abolish it . . . .

"Democratic clubs" sprang up throughout the South. Citizens banded together to contest the corruption that had consumed state and local governments. Goals of such clubs were (1) ouster of scalawags and carpetbaggers from government, (2) restoration of voting rights for all men, and (3) restoration of constitutional liberty. On 2 May 1868 the new Dorn's Mine Democratic Club prepared a constitution which reads in part:

> Believing the people of our state *[to be]* firmly attached to the principles of constitutional liberties as handed down to us by the fathers of the Republic, and recognizing the National Democratic Party as the only party true to those principles . . . and being anxious to preserve peace, law, and order by uniting for the overthrow of the enemies of our

> peace and country; therefore be it RESOLVED, that we form ourselves into a Democratic Club to be known as the Dorn's Mine Democratic Club; *[and]* RESOLVED, that those Negro men who voted the Democratic ticket or abstained from voting, we recognize as our friends, and *[we]* will stand by them as such and give them all the assistance in our power, and to those who have acted against us hitherto, we now extend an invitation to join us in the name of liberty; *[and]* RESOLVED, that we *[will]* hold as enemies all *[who]* will not unite with us . . . .

Seven days later on 9 May other frustrated citizens formed the Democratic Club of Liberty Hill. Club members formalized their club by adopting a charter which reads in part:

> Being anxious to maintain and preserve law and order and to promote the welfare of all classes of people, *[we]* are ready and prepared to cooperate with the *[Democrats]* of our state, and of the United States, in any line of policy that will be constant with those great principles of liberty; *[and]* RESOLVED, therefore, that we organize ourselves into a club known as the Democratic Club of Liberty Hill . . . and earnestly solicit all good citizens to unite with us in arresting the dangers which threaten our Constitutional liberty; *[and]* RESOLVED, that the duties of the members of this club shall be to use their utmost influence and exertion in peaceable opposition to any measure which, in their judgement, is in violation of the Constitution of the United States.

Rise of Vigilante Organizations: Not content with the posture of new democratic clubs, many disenfranchised southerners banded together in "secret societies" and vigilante groups such as:

> White League: This coalition of rag-tag militias first formed in Louisiana, and it rapidly spread throughout the South. Its goals were twofold: (1) violent overthrow of Radical Republican tyranny, and (2) terrorization of Republican voters and sympathizers.

Klu Klux Klan: First formed in Tennessee by a small group of Confederate Army veterans, the Klan spread like wildfire into other southern states. It would become the most violent group opposing Radical Republican rule.

Red Shirts: This paramilitary group formed in the Carolinas with the goal of preventing people from voting for Radical candidates. In the next decade Red Shirts would be the deciding factor in the crucial 1876 national election.

Rifle Clubs and Sabre Clubs: These clubs were independent groups of armed vigilantes. There was little or no connection between them. They mimicked many activities of the White League, the Klu Klux Klan, and the Red Shirts.

The most violent vigilante group proved to be the Klu Klux Klan. On Christmas Eve in 1865, six Confederate Army veterans had formed this secret society in Pulaski, Tennessee. Similar "klaverns" soon formed in all other southern states, but beyond the local level there was little organizational structure. Klan goals were: (1) ouster of occupying U.S. Army troops, (2) overthrow of Radical Republican government, and (3) restoration of democratic principles. However, despite their pseudo-noble stated principles, Klan members were armed insurgents and terrorists. Many Klansmen were educated members of the former antebellum society, but others came from diverse backgrounds described by one historian:

> Lifting the Klan mask revealed a chaotic multitude of vigilante groups, disgruntled poor farmers, wartime guerrilla bands, displaced Democratic officeholders, illegal whiskey distillers, coercive moral reformers, workmen fearful of competition, employers trying to enforce labor discipline, and even a few Negro freedmen and Republicans who had agendas of their own. All they had in common was that they called themselves Klansmen.

Klan members wearing white robes, conical hats, and masks to conceal their identity roamed the countryside on horseback at night. They burned hundreds of homes, sometimes with residents still inside, and they intimidated or murdered Republicans and freedmen who supported the Radical regime. The Klan served as the unofficial paramilitary arm of the Democratic Party in the South. In general, Klan members were oath-bound to support *The Prescript*, the Klan dogma. The lengthy creed and preamble are quoted below in part:

> We, the Order of the ____________________, reverently acknowledge the majesty and supremacy of the Divine Being and recognize the goodness and providence of the same. And we recognize our relation to the United States Government, the supremacy of the Constitution, the Constitutional laws thereunder, and the Union of the States thereunder . . . .
> *[The Prescript, eventually formalized at Nashville, Tennessee, in April 1867]*

Former Confederate Gen. Nathan Bedford Forrest assumed the title of Grand Wizard of the Klan and verbally claimed to be the Klan's national leader. Yet, there were no national, state, or regional offices. Klan units operated autonomously, and in the countryside at night the Klan prowled at will. Southern states degenerated into outright guerrilla warfare, pure and simple. Thousands were shot, hanged, drowned, burned, or beaten to death.

To combat the insurgency Ulysses Grant, who had become the U.S. President in 1868, suspended habeas corpus and sent more and more troops to the South. Nonetheless the insurgency continued. One Republican politician desperately pleaded for more military help from Washington and wrote:

> Armed guerrilla warfare killed thousands of *[Republican supporters]*. Political riots were staged. Their causes were always obscure; their results were always certain, ten to one-hundred times as many Republicans killed as Democrats killed. The *[U.S.]* Army is out-gunned and out-manned. Our vaunted state militia cowers indoors at night and does nothing to stop the slaughter.

Violence was not always necessary to induce compliance with Klan demands. The mere threat, real or implied, of violence often had the same effect. The narrative below describes the awesome persuasive power of the robed and masked night-riders. After midnight they went to the rural home of a female schoolteacher:

> *[She]* was visited by the Klan between one and two o'clock in the morning in March 1871 by about fifty men, all mounted and disguised. Each man wore a long white robe, and each face was covered by a loose mask with scarlet stripes. She was ordered to get up and dress, which she did at once. She then admitted to her room the *[Klan]* captain and lieutenant . . . . Eight or ten Klansmen stood inside the door, and the porch was full *[with more Klansmen]*. They treated her gentlemanly and quietly, but complained of the heavy school tax. *[They]* said she must stop teaching and go away. They warned her that they never gave a second notice. She heeded the warning, left the county, and never returned.

The South suffered terrorism on an almost unimaginable scale. In addition to the Klan, citizens began forming rifle clubs and saber clubs. These armed vigilante bands conducted a reign of terror against Republican leaders of both races. They attacked or intimidated thousands. In 1871 the desperate U.S. President ordered all rifle and saber clubs to disband immediately and surrender their firearms. In response the clubs simply changed their names. The infamous Allendale Rifle Club morphed overnight into the Allendale Mounted Baseball Club. Saber clubs suddenly became horseshoe clubs, fishing clubs, and missionary societies. Nothing had changed except the names. Vigilantes prowled at will. The killing continued, and more thousands succumbed to the violence.

On Easter Sunday, 13 April 1873, vigilantes faced the state militia at Colfax Courthouse in rural Grant Parish, Louisiana. No one now knows who fired the first shots, but mayhem erupted. When the smoke cleared about 142 men had been killed. That total includes 40 Republican functionaries who initially were captured, but who were lined up and shot several hours later.

During 1874 in Louisiana open warfare erupted in the city of New Orleans. In what became known as the Battle of Liberty Place, over 5,000 White League members charged into the city and confronted a combined force of police and state militiamen. After battling and routing the state forces the White League occupied federal offices for three days. They left the city when hundreds of U.S. Army troops arrived and threatened to attack them.

In North Carolina an estimated 22,000 men had joined the Red Shirts, which they called the "military arm of democracy." Overt warfare raged in southern states. Armed bands, by whatever name, struggled to oust occupying U.S. Army troops and restore democratic and representative government. Radical Republican officeholders appealed to President Grant for more military support. Frustrated by violence that was growing worse and worse, Grant refused to act. He explained that the northern public was "tired out" of the perpetual violence in the South.

Vigilante Justice?: Most vigilante violence was aimed at corruption in government, but there were exceptions. In the rural Winter Seat community a band of robbers brutally murdered an old man and his wife. The animalistic thugs heinously beat the elderly couple to death in their home and then, for good measure, slit their throats. The *Edgefield Advertiser* newspaper opined on 31 May 1876:

> The annals of crime will scarcely furnish a parallel to the murder of Mr. and Mrs. John L. Harmon at their home in Winter Seat.

Eight culprits, six men and two women, were caught. Several of them confessed to the crime and implicated all the others. With no trust in the Radical Republican judicial system, vigilantes struck. They abducted the sheriff, marched the six murderers out of town, gave them several minutes to pray, and then shot them. As a gesture of good will they released the two female accomplices.

News of this "vigilante justice" spread nationwide. Almost all newspaper editors in the North had grown weary of the continuing corruption and bloodshed in the South. They resented the continuing financial expense of maintaining thousands of soldiers there to prop up a government that no longer represented the average citizen. The northern public sympathized with the vigilantes, and the *New York World* newspaper editorialized:

> The people at last grew tired of dilatory or corrupt courts, insecure jails, and easily obtained pardons, and when the Harmons – old, feeble, and inoffensive folks – were knocked on the head and had their throats cut by a gang of

> murderers, the *[vigilantes]* took the scoundrels and put them to death.

The Hamburg Riot: By early 1876 emboldened vigilante bands had begun making armed forays into cities in plain view of witnesses during daylight. In downtown Hamburg (later, *North Augusta*), South Carolina, across the Savannah River from Augusta, Georgia, violence would erupt on 8 July 1876. A wagon on Market Street had been blocked by armed state militiamen. Harsh angry words had been exchanged between Democrats in the wagon and Radical Republicans in the state militia.

Hundreds of vigilantes stormed into Hamburg. The outgunned militia captain and about 38 of his men took refuge in the local armory. Vigilantes fired at the armory, but militiamen escaped through a back door. Vigilantes soon found 25 of them who had been hiding in crawl spaces under houses in the town. One tried to get away, and vigilantes shot him. After a quick kangaroo court they shot five more captured militiamen.

News of the cold-blooded murder of captured militiamen, in daylight and in front of dozens of witnesses, spread via telegraph and newspaper throughout the United States. The killings electrified the nation. The highly publicized riot and murders in Hamburg would have a dramatic effect on the national election four months later.

The Critical National Election in 1876: Throughout the South the possibility of change was in the wind during autumn months of 1876. Terrorist groups continued to target and chip away at the Radical Republican establishment. Federal troops could do little to stem violence in the cities, and they had no control in the countryside. Yet, misuse and theft of public funds continued. Historian Walter Edgar explained in writing:

> The administration of *[governor]* Robert K. Scott was so corrupt that a group of Republican legislators *[from his own party]* tried to impeach him and treasurer Niles G. Parker in 1872. The effort failed after the governor bribed enough legislators to get off scot-free. . . . Franklin J. Moss

> Jr. succeeded Scott and carried the graft and corruption to new levels.

Many Negro freemen, angry that Radical Republican government had not helped them get land, began to retaliate. They changed allegiances and supported the Democrats. Meanwhile the Democrats used violence, intimidation, and outright fraud in an effort to wrest state governments away from Republicans. Red Shirts, fully armed, rode to all Republican rallies and demanded the right to speak. The Radical establishment fought back with its judicial system, and many Red Shirts were thrown into jail.

Election day arrived on 3 November 1876. It was marked by fraud and intimidation on both sides. Voting was one thing, but having the votes *counted* was another matter. Armed vigilantes surrounded many polling places, and the result was obvious. In South Carolina after all votes were tallied, Laurens and Edgefield Counties had more Democratic votes than the entire number of registered voters. Republicans naturally cried foul, and the state legislature voided the election results.

On the national level in the race for the presidency the election had pitted Samuel Tilden, a Republican, against Rutherford Hayes, a Democrat. Just like the dilemma in South Carolina, the national vote tally also was in dispute, and Congress had to sort out the madness. President Grant signed the Electoral Commission Act. It set up a 15-member group of congressmen to settle the disputed presidential election results.

In smoke-filled rooms and with street-corner bargains, Hayes advocates garnered support. In return for a convoluted victory for Hayes, congressmen and senators agreed that they would (1) declare Wade Hampton, a Democrat, to be the gubernatorial victor in South Carolina, and (2) Hayes would withdraw occupying U.S. Army troops from the South.

Thus, in this "Bargain of 1877" the U.S. Congress officially declared Rutherford Hayes to be the presidential election winner by one Electoral College vote. As the new commander-in-chief of the armed forces, Hayes remained true to the agreement. After taking office on 4 March 1877 he ordered all occupying U.S. Army troops to leave southern states. One embittered southern governor, Daniel

Chamberlain, drafted a proclamation to his staff and to citizens who had supported his Radical Republican agenda:

> The government of the United States abandons you, deliberately withdraws from you its support, with full knowledge that the lawful government of the State will be speedily overthrown. . . . Lawful State Government *[has been overwhelmed in]* a struggle with insurrectionary forces too powerful to be resisted.

The Reconstruction Era Ends: Republican control collapsed in the South. For over ten years opportunists and profiteers had controlled southern state governments at all levels. Yet, in the end corruption proved to be their Achilles' heel. Republican officeholders who had not been defeated at the ballot box either resigned or were drummed out of office. Their opportunity to fleece the public had come to an end. Within months southern state governments and judiciaries all were controlled by Democrats. As agreed, occupying U.S. Army troops left the South, and new officeholders restored voting rights to those who had been disenfranchised. Walter Edgar would write:

> *[Southern states]* were able to regain control . . . because they mounted an insurgency against what they viewed as an alien government imposed upon them by a conquering foe. From 1865 to 1877 they successfully prosecuted what today would be called an insurgency, a people's war . . . .

Despite the end of the scalawag and carpetbagger regime there was little else to celebrate. Over a decade of corruption and graft had taken a fearsome toll. The infrastructure and economy of the South had degenerated into ruin. The entire region would have to be rebuilt, a task that would take most of a century.

The Human Toll: The human toll cannot be erased. The American Civil War officially had ended in April 1865 when the C.S. Army capitulated. Yet, in a sense the war had continued for another 12 years, renamed as the Reconstruction Era. Untold tens-of-thousands had been killed. Some had been beaten to death, others had been shot, some had been hanged, and occasionally some had been burned alive in their homes.

Carnage during the Reconstruction Era consumed innocent people of both sexes, both races, all ages, and all political persuasions. Men, women, and children perished by the thousands. Yet, one must add their unborn posterity to the death toll. Those unborn children, grand-children, and great-grandchildren had been *killed* as well, for they never lived to see the light of day. They all must be included when tallying casualties during the Reconstruction Era.

> Every generation is doomed to fight its war, to endure the same old experiences, suffer the loss of the same old illusions . . . They were fighting in the cruelest kind of conflict, a war for survival waged . . . without rules or laws; each man fought for his own life and the lives of *[those]* beside him, not caring who he killed, or how many, or in what manner, and feeling only contempt for those who sought to impose on his savage struggle the mincing distinctions of civilized warfare.
> *[Philip Caputo, A Rumor of War, 1977]*

Etching of a carpetbagger (from ***Harper's Weekly*** *in 1868).*

Etching depicting White League members attacking police and militia in New Orleans (from ***Harper's Weekly*** *in 1874).*

Theatrical poster advertising The Birth of a Nation, a three-hour silent film released in 1915. The film chronicled the lives of two families during the American Civil War and the Reconstruction Era. The film glorified the Klu Klux Klan and was highly controversial because of its racial emphasis. It was the first motion picture to be screened in the White House.

Image of the advertisement logo for the Confederada Festival held each year in Americana, Brazil (the logo is written in Portugese, the common language spoken there). The logo was dated for the festival on 30 April 2017. Each year the festival features dancing, barbeque, Brazilian food, and elaborate costumes for men and women. Flags of the United States, the former Confederate States, and Brazil pay homage to the mixed heritage of community founders and their descendants. Although English is still spoken as a second language by some elderly Americana residents, fifth generation youngsters have fully blended into modern Brazilian society.

> Sure, I wave the American flag. Do you know of a better flag to wave? Sure, I love my country with all her faults. I'm not ashamed of that. Never have been. Never will be.
> *[John Wayne, aka Marion Morrison (1907-1979)]*

Casualties of War

> The muster rolls on which the name and oath were written were pledges of honor, redeemable at the gates of death. And those who went up to them, knowing this, are on the list of heroes.
> *[BGen. Joshua L. Chamberlain, USA, writing about Union and Confederate volunteers, 1866]*

Death Toll during the American Civil War: During the American Civil War the human toll devastated America. Many thousands of soldiers had been killed in battle or had died from wounds, disease, and privation. A modern-day consensus reveals that the war snuffed out the lives of at least 741,000 soldiers. In addition to those who were killed, about 325,000 more had been crippled, dismembered, blinded, or otherwise maimed and incapacitated. The war created the greatest human catastrophe in American history.

In the modern era if Americans think of bloody warfare, most of them reflect upon the two world wars. Those wars featured carnage on a massive scale. Yet, their military death toll for Americans was dwarfed by the American Civil War. In fact, military deaths during the American Civil War exceed United States military deaths from all other wars combined.

Both the U.S. Department of Defense and the U.S. Department of Veterans Affairs maintain military death records for major American wars (death totals vary, for accounting methods differ). These records of military deaths are listed chronologically below:

Common Name	Military deaths per D.O.D.	Military deaths per V.A.
Indian Wars	(no estimate)	1,000 (est.)
Revolutionary War	(no estimate)	4,435
War of 1812	(no estimate)	2,260
Mexican War	13,283	13,283
American Civil War	*see below*	*see below*
Spanish-American War	2,446	2,446
World War I	116,516	116,516

World War II.	405,399..	405,399
Korean War.	36,516..	36,574
Vietnam War.	58,209..	58,220
Gulf War.	258.	383
Middle East & Afghanistan.	6,884..	6,782
Totals:.	639,511..	647,298

American Civil War estimated military deaths . . .741,000

If prorated to parallel United States population growth over the past 150 years, American Civil War military deaths will stagger the senses. The 1860 Census was completed in November of that year, and the United States population then totaled 31,443,321 persons. One-hundred-fifty years later the 2010 Census tallied 308,745,538 persons, an almost ten-fold increase of 980 percent. Therefore, if the 741,000 American Civil War military deaths are prorated to account for population growth from 1860 to 2010, the death toll rockets up to 7,261,000 persons – over seven million deaths.

> The overall mortality rate for the South exceeded that of any country in World War I and all but the region between the Rhine and the Volga *[Rivers]* in World War II.
> *[James McPherson (1936 —), Civil War historian]*

The national death rate during the American Civil War, adjusted for the population increase by the 1940s, was six times higher than the death rate during World War II. And because of the regional population difference, Confederate soldiers were four times more likely to die in battle than their Union counterparts. The war snuffed out lives of healthy young soldiers at a fearful rate, and it did not spare civilians.

"Bleeding Kansas" was an accurate moniker. In the wild west border states of Missouri, Kentucky, and Kansas there was little distinction between soldiers and civilians. Bands of partisans and guerrillas slaughtered all who disagreed with them. An estimated 54,000 non-combatants of all ages and both sexes perished.

How Many *Really* Died?: In the 1890s, 30 years after the war ended, scholars made the first serious attempts to determine how many lives had been lost during the war. It proved to be a daunting task. Neither the North nor the South had maintained cumulative death tallies. Neither side maintained what later would be known as standardized personnel records. There had been no modern-day unit diaries or command chronologies. Roughly half of all soldiers killed in battle had been unidentified. After-action reports, if they could be found, sometimes listed men killed in combat. At other times there were no such listings. Rosters and muster rolls of military units in the North and South had been designed primarily for verification of ready-for-action troop counts.

After years of periodic research, scholars collectively decided upon an estimate of 620,000 military deaths. Once that figure was published it became set in stone. Why question it?

> That traditional estimate had become iconic. It had been quoted for the last hundred years or more.
> *[Rachel Coker, historian, writing in 2011]*

Recent analysis has shown that the old 620,000 tally is far too low. David Hacker, an ardent demographic historian at Binghamton University, headed a recent study. He researched not only the growing number of on-line records, but he also delved into 1860 and 1870 census data. Also he researched pension applications from war widows. His final tally recorded "at least 750,000 military deaths." Much of this 130,000 increase was due to previously unidentified Confederate losses.

The new tally includes soldiers absent on wound furlough who succumbed to their wound at home. It also includes the estimated 56,000 who died in squalid POW camps. However, the war had been fought at the beginning of modern statistical concepts and record-keeping. A precise death toll never will be known.

> The new death count is greater than all American war deaths from other conflicts combined, from the Revolution to the present day.
> *[Rachel Coker, historian, writing in 2011]*

Yet, Hacker's estimate does not include the estimated 54,000 civilians who were shot, hanged, drowned, beaten to death, or burned to death during the war. It also does not include the 325,000 soldiers who survived but were occupationally disabled. These maimed men had fallen prey to crippling combat injuries (63,000 of them had lost at least one limb), blindness, camp diseases, and assorted physical trauma. They all must be added to the list of casualties of war.

A Price Paid in Blood: The human toll of the war fell much more heavily on the South than on the North. The northern population was about 22 million, but only 5 million freemen lived in the South. Southern military losses had to be absorbed by a population only 23 percent that of the North. When southern and northern casualties are viewed together the totals almost defy comprehension.

> The *[51,000]* casualties suffered in a single day at the Battle of Antietam *[aka, Battle of Sharpsburg]* on September 17, 1862, was four times the number of Americans killed and wounded at the Normandy beaches on D-Day, June 6, 1944. More Americans were killed in action that September day near Sharpsburg, Maryland, than died in combat in all the other wars fought by the United States in the 19th Century.
> *[James McPherson, Out of War, a New Nation, 2010]*

Confederate soldiers paid a dear price. The Confederacy kept no master records to document the number of wounded survivors, or those who were disabled or crippled. Yet, Confederate documents reveal that over 26,000 southern survivors had at least one limb amputated. More thousands were otherwise maimed for life. For many of these men, especially those who had lost more than one arm or leg, post-war labor on farms was improbable or impossible. As a result many families in the South became destitute.

Union Army records show that 214,000 Confederate soldiers were confined in prisoner of war camps at one time or another. Of those, slightly over 26,000 (12 percent) died in captivity.

The three Briggs brothers (see the first chapter, "The Incalculable Toll of War") had marched off to war. They all died, and their entire family was swept away. However, such total losses were not the norm. The experience of the Briggs brothers' neighbors was more typical in the South. All four brothers in the neighboring farm family in western South Carolina had joined the C.S. Army:

– One was killed-in-action at Dandridge, Tennessee.

– One was wounded, captured, and imprisoned in a POW camp.
– One was wounded by artillery fire and maimed for life.
– One survived the war unscathed.

Battle Casualty Rates: Alfred Tennyson had immortalized an ill-fated cavalry charge in his poem, *The Charge of the Light Brigade.* The poem refers to the Battle of Balaklava in 1854 during the Crimean War. About 676 British light cavalrymen had charged into a valley to attack Russian artillery and its crews. The artillery crews were supported by Russian infantry battalions. The charge reached the Russian batteries, but 278 of the cavalrymen (41 percent) became casualties. Surviving British cavalrymen were forced to retreat.

British casualties at Balaklava were far surpassed by southern losses on many occasions during the American Civil War. During three days of fighting at Gettysburg on 1-3 July 1863, the 26th North Carolina Infantry Regiment lost 714 of its 800 soldiers (89 percent casualties). Both in terms of percentage casualties and number of casualties, it was the largest regimental loss in a single battle during the war. On the first day alone the regiment lost 584 men. When the roll was called for Company G the next morning only one man was present, and he was wounded.

Many other Confederate units also suffered horrible losses. In March 1865, one month before the end of the war, Company B of the Irish Brigade from Louisiana began the day with 98 men. At the end of the day only two were left. Company A began with 91 men, and the next morning only four were left. Some heavy regimental losses in single-day battles – *only one day* – are listed below:

Regiment	Battle	Casualties
1st Texas Regiment.	Sharpsburg.	82 percent
21st Georgia Regiment..	Bull Run.	76 percent
6th Mississippi Regiment.. . .	Shiloh.	71 percent
8th Tennessee Regiment. . . .	Stone's River. . . .	69 percent
10th Tennessee Regiment. . .	Chickamauga. . . .	68 percent
Palmetto Sharpshooters.	Frayser's Farm. . .	68 percent

On 18 September 1862, the morning after the terrible slaughter at Sharpsburg, Maryland, a young Confederate soldier peeped out of his trench and looked at the field of battle. He wrote in his diary:

> The sun rose to a sickening sight. Hundreds of corpses lay on the field and, in places, were piled in rows where soldiers had been cut down as they advanced in formation. Untold numbers of injured troops were still crying for help. Only a few hundred yards separated the two armies. If a soldier ventured onto this no-man's-land, a fusillade of musket fire greeted his appearance, *[so]* the dead and wounded *[had to]* go unattended.

The Cost of Disability: Those who try to romanticize war are either naive, or they reject reality. War is about exhaustion, filth, disease, privation, agony, and horror. In winter it is about trying to sleep on the ground in sub-freezing weather, sometimes in the rain or snow, with no shelter or blanket. In summer it is about living among swarms of flying, buzzing, crawling, stinging, sucking insects that make days miserable and nights unbearable. Over a century later in 1981, author Mark Baker would explain:

> War is not about killing. Killing is the easiest part of the whole thing.

Wounded and disabled Confederate soldiers straggled homeward after the war. Many of them found that their farms had been sold or abandoned, and their impoverished families had scattered. Somehow these debilitated survivors had to become self-sufficient or at least obtain food and shelter.

Amputees' needs were the most visible, and one out of each 13 survivors had lost at least one limb. From limited resources Georgia and Louisiana provided $100 per year to each war amputee. In 1866, the year after the war ended, Louisiana budgeted $20,000 to buy prosthetic arms and legs for its veterans. Tennessee gave a ten dollar per month stipend to veterans who had been blinded "while engaged in battle."

South Carolina authorized pensions for war veterans in 1887. Pensioners had to have a disability that rendered them unable to make a living, and those who met that criteria got five dollars per month. In later years the criteria for recipients was loosened, and monthly payments increased.

Some cash-strapped states provided land grants instead of money.

Louisiana granted 123,000 acres of "swamp land" (of dubious value) to disabled and indigent veterans. Texas could afford to be more generous with land grants and gave 1,979,000 acres to handicapped war veterans.

Most states in the South eventually began supplying prosthetic limbs to war amputees. One major supplier, Dr. Douglas Bly from Rochester, New York, offered his "Anatomical Patent Leg" to states for $150. He also marketed his more economical "Army and Navy Leg" for $74. By the late 1870s states began paying not only for prosthetic legs, but prosthetic arms as well. By the 1880s some states began making modest disability payments to veterans who had lost an eye during the war.

As years passed states could afford to be more generous. In the 1890s decade Georgia spent an average of $500,000 per year on pensions. During some years the pension payments exceeded ten percent of the entire state budget. Texas levied a special tax to fund pensions for indigent and disabled veterans. One historian noted:

> By the turn of the century Alabama was appropriating $250,000 annually for *[Confederate]* pensions. The state constitution mandated that ten cents out of every dollar in revenue *[must]* go toward funding pensions.

Wounded and Disabled Soldiers: Despite state efforts to assist disabled veterans, one type of disability had been largely ignored during and after the war. Wounds received in battle sometimes extended far beyond physical injuries. For example, a half-century later during the Great War (aka, *World War I*) in Europe, medical professionals would coin the term War Neurosis. The public would call it shell-shock. Another half-century later medical personnel would begin using a new term for extreme psychological injuries in warfare, Post Traumatic Stress Disorder.

After the terrors of battle or the horrors of POW camps, some soldiers suffered a type of nervous system disorder that transformed them into shaking and convulsing wrecks. Some became catatonic and retreated into mental oblivion, unable to function usefully in society. Often such mental wounds failed to heal with the passage of time. When these mental injuries combined with visible physical wounds, the victim often was left totally dependent upon others for shelter, care, and sustenance. Thousands of incapacitated soldiers who were mentally injured or physically injured, or both, are among the many casualties of war. Tragically these human costs are not one-time losses. They continue and accrue year after year.

Confederate Soldiers' Homes: By 1890, 25 years after the war had ended, half of all southern veterans already had died. The federal census that year enumerated 428,747 living war veterans in the South. Because of war wounds and injuries many were disabled, and their care fell to families, churches, and communities.

Five years earlier in 1885 the first permanent group home, Lee Camp Soldiers Home, had opened in Richmond, Virginia. Union veterans made contributions and called the home "a monument to a reunited country." Over the years 15 more similar homes opened (ironically the last one opened in 1929 in Los Angeles, California). Criteria for acceptance were rigid. Residents had to be indigent and unable to care for themselves, must have an honorable discharge or parole, and must be able to prove war-related disability.

No combined records remain, but about 16,000 infirm veterans entered the homes at one time or another. The homes, often called asylums, were a refuge, a military camp, a small community, and a shrine. To the public they were a living reminder of the past, a symbol of sacrifice to be revered.

The Confederate Soldiers Home in South Carolina was typical of most others. It had been authorized by the state legislature, and it opened its doors to residents on 10 May 1909. Two veterans from each county had been approved by county pension boards. Residents operated a 24-acre farm to supplement support from the state. Sixteen years later in 1925 one indigent Confederate widow or wife per county was eligible for admittance. "Sisters of full blood" became eligible in 1929, and daughters gained eligibility in 1935. Beginning in 1943, indigent nieces could be admitted.

The last male Confederate veteran in the home died in 1944, but the home remained open 13 more years. It closed in 1957, and its 29 elderly women residents were transferred to the State Department of Public Welfare. The structure was demolished six years later. A granite monument on the site is inscribed:

CONFEDERATE SOLDIERS
HOME
By an act of the General Assembly
in 1908, an infirmary was established
on this site for the infirm and
destitute Confederate Soldiers and
Sailors of this state. In 1925
eligibility for admission was
extended to wives and widows of
Confederate Veterans. The home was
closed in 1957, when too few
residents remained.

Memorial Day in the North and South: During and after the war Americans often placed flowers on graves of combatants who had fallen in battle. On 25 April 1866, the year after the war had ended, women in Mississippi went to a cemetery and laid flowers on graves of Confederate soldiers killed in 1862 during the Battle of Shiloh in Tennessee. They noticed bare and neglected Union graves nearby, so they placed flowers on those graves as well.

Two years later in March 1868 at Columbus, Georgia, a military widow laid flowers on her husband's grave. She and her friends also laid flowers on graves of other men who had been killed during the war. The widow wrote to the *Columbus Dispatch* newspaper:

> . . . We beg the assistance of the press and the ladies throughout the South to aid us in the effort to set apart a certain day to be observed from the Potomac to the Rio Grande, and to be handed down through time as a religious custom of the South, to wreathe the graves of our martyred dead with flowers, and we propose the 26th day of April as the day.

The request prompted action not only in southern states, but also in the North. The next month on 5 May, Gen. John Logan, commander of the Grand Army of the Potomac (a Union veterans organization), issued General Order No. 11. The order designated 30 May as a day for "strewing with flowers" the graves of soldiers who had been killed during the war. On 30 May in Arlington National

Cemetery citizens offered prayers, sang hymns, and placed flowers on graves of Union and Confederate soldiers.

This practice spread across America, and 30 May became known as Decoration Day. That term soon was replaced with a new name, Memorial Day. A century later in 1967, Congress finally made it a national holiday. Per federal law, Memorial Day is set aside to honor those from American armed forces who lost their lives in the service of their country during wars or armed conflicts. Citizens recently have been asked to fly the POW/MIA flag approved under the 1998 Defense Authorization Act, observe a "moment of remembrance" at 3:00 p.m., and pledge to aid families of the "honored dead."

In addition to the national holiday 12 southern states observe a state holiday, Confederate Memorial Day. On this day they honor men of the Confederate States armed forces who died in the service of their country. One of the 12 states, Kentucky, was a border state and never joined the Confederacy. Nonetheless, its state legislature recognizes and observes Confederate Memorial Day. Each of the 12 states has selected its own day of observance as follows:

State	Day of Observance
Alabama	Fourth Monday in April
Arkansas	19 January (birthday of Robert E. Lee)
Florida	26 April (day of Gen. Johnson's surrender)
Georgia	26 April (day of Gen. Johnson's surrender)
Kentucky	3 June (birthday of Jefferson Davis)
Louisiana	3 June (birthday of Jefferson Davis)
Mississippi	Fourth Monday in April
North Carolina	10 May (day of Gen. Jackson's death)
South Carolina	10 May (day of Gen. Jackson's death)
Tennessee	3 June (birthday of Jefferson Davis)
Texas	19 January (birthday of Robert E. Lee)
Virginia	Last Monday in May

Tomb of the Unknown Confederate Soldier: After World War I ended the British government chose one of the many unidentified British soldiers who had been killed in France. With King George V attending, the body was interred in Westminster Abbey on Armistice Day in 1920. A similar gesture of honor took place in France when an unidentified French soldier was interred with military honors at the

Arc de Triomphe.

The next year on Armistice Day in 1921 the United States interred an unidentified American combatant in Arlington National Cemetery. The tomb never has been officially named, but usually it is called the Tomb of the Unknown Soldier. The large white marble sarcophagus is inscribed: "Here Rests In Honored Glory An American Soldier Known But To God."

Three decades later in Philadelphia, workmen chanced to unearth remains of a Revolutionary War soldier. He was identified by metal remnants of his uniform. A skeletal examination showed that he had been killed by a musket ball. In 1956 the soldier's remains were interred in the Tomb of the Unknown Revolutionary War Soldier in Washington Square in Philadelphia. The lengthy tomb inscription and epitaph begin: "Beneath this stone rests a soldier of Washington's Army who died to give you liberty."

In 1979 at an old battlefield site near Vicksburg, Mississippi, workmen stumbled upon remains of an unidentified Confederate soldier. Several organizations including (1) Sons of Confederate Veterans, (2) United Daughters of the Confederacy, and (3) Military Order of the Stars and Bars began a campaign to honor the fallen soldier and other unidentified Confederate war dead. Two years later on 6 June 1981 the unidentified solider was interred in the Tomb of the Unknown Confederate Soldier in Biloxi, Mississippi, at Beauvoir (meaning, *beautiful view*), the Jefferson Davis Home and Presidential Library. The tomb bears the Great Seal of the Confederacy and the following inscription:

The Unknown Soldier
of the
Confederate States of America
Known But To God

During the war the C.S. Congress had authorized medals to honor valor in battle. Criteria for the C.S. Medal of Honor were almost identical to requirements for the U.S. Medal of Honor: intrepidity, gallantry, and risk of one's life above and beyond the call of duty. Although approved, none had been awarded during the war.

During a formal ceremony in May 1983 the Sons of Confederate Veterans organization posthumously awarded the C.S. Medal of

Honor to the Unknown Confederate Soldier. The medal remains on display inside the museum at Beauvoir.

Parting Shots: Time gradually took its mortal toll on veterans of the American Civil War and their families. Per the U.S. Veterans Administration, the last *verified* living veteran was Albert Woolson. He died on 2 August 1956 at 109 years of age.

The last *verified* living veteran's widow was Maudie Hopkins. She died on 1 August 2008 at 93 years of age.

> War is an unmixed evil of blood, butchery, desolation, death, robbery, rapine, and violence – palliated only when waged in self-defense.
> *[a Mississippi cavalryman, in a letter to his family]*

Some of the Confederate dead scattered along the Hagerstown Turnpike near Sharpsburg, Maryland, on 17 September 1862.

Some of the Confederate dead at Dunker Church near Sharpsburg, Maryland, on 17 September 1862.

Some of the Union dead lying on the battlefield near Gettysburg, Pennsylvania, after the struggle on 1-3 July 1863.

Some of the Confederate dead lying behind the protective stone wall at Marye's Heights on 3 May 1863.

Three convalescing amputees.

Building a New Nation

> These dead, from the North and from the South, belong to all of us. The last hateful memory that could divide our country is buried with them. About their graves kneels a new nation.
> *[Historical Committee, United Confederate Veterans]*

A Devastated Society: As the war sputtered to a merciful end in April 1865, people in the South faced a bleak and uncertain future. Four years of war had saved the union of states, but the toll of that salvation had devastated the former Confederacy. Union armies had left destruction in their wake. Like Atlanta, many southern cities had been burned to the ground. There was no money with which to rebuild. Farmers had no seed with which to replant. The economy and infrastructure lay in shambles. Food was scarce, and the horrible specter of starvation faced many families.

Against this backdrop some former Confederate soldiers talked openly of continuing the fight. Guerrilla warfare might be their best option, they reasoned. Former Confederate commanders disagreed. Gen. Robert E. Lee famously had admonished a mother to reconcile with the Union and "make your sons Americans." Gen. Nathan B. Forrest stressed unity with the United States. In a farewell address to his soldiers he wrote:

> It is a clear fact that we are beaten. We would be foolish to try to fight further. The government which we tried to establish is at an end. Civil war – such as you have passed through – naturally causes feelings of bitterness and hatred. We must put those feelings aside. Whatever your responsibilities may be, meet them like men. You have been good soldiers. You now can be good citizens.

Blue-Gray Reunions: There is a time-worn adage often voiced by war veterans in reference to civilian *armchair warriors*: "Hell hath no fury like the fury of a non-combatant." Warfare is not for the faint of heart, and dreams of glory are illusionary. Those who have survived close-combat know the searing emotions involved. In close-combat a man is face to face with the enemy. He can look into the eyes of the man who is trying to kill him. It is an incommunicable experience. Late in life, survivors who endured horrors of battle often espouse high regard, and even admiration, for their former enemies. Unlike civilians, they can appreciate the experience of meeting men who once were their fearsome adversaries, stalwart men who shared the same perils and hardships.

Veterans of the war in the North and South began to advocate unity and forgiveness by honoring bravery and devotion to cause. They made no attempt to glorify the bloody struggle. Instead they sought common ground in honorable service, loyalty, and a saved country. They were among the first to promote the new nationalism that swept through the United States after the war.

Within a few years after the war ended former Confederate and Union soldiers who survived began gathering for small Blue-Gray reunions. The first such reunion of old adversaries to attract over 1,000 veterans was held at Bunker Hill, Virginia, in 1875, ten years after the war ended. That event was well publicized in newspapers, and it opened the floodgates for many more such reunions. Blue-Gray memorial services became common. While attending Mardi Gras in New Orleans in 1881 a Union veterans group held a memorial service and placed flowers on graves of Confederate soldiers. The next year, 1882, former adversaries held a huge reunion at Gettysburg battlefield in Pennsylvania.

Most civilians cannot fathom why, years after a war, combat veterans feel compelled to meet wartime comrades and adversaries. Why not forget about the war and move forward with the business of living? Perhaps that question was best answered many years later by a combat veteran, Michael Norman. With regard to aging wartime comrades, he explained in *These Good Men*:

> I now know why men who have been to war yearn to reunite. Not to tell stories or look at old pictures. Not to laugh or weep. Comrades gather because they long to be with men who once acted their best, men who suffered and sacrificed, who were stripped of their humanity. . . . I know them in a way I know no other men. I have never given anyone such trust. They were willing to guard something more precious than my life. They would have carried my

> reputation, the memory of me. It was part of the bargain we all made, the reason we were willing to die for one another. As long as I have memory I will think of them all, every day. I am sure that when I leave this world my last thoughts will be of my family and my comrades – such good men.

About 10,000 Union veterans were joined by 3,000 Confederate veterans at a gathering in 1887 at Evansville, Indiana. The Indiana governor addressed the crowd and remarked: "The issues upon which the *[war]* was based are now buried." That same year Confederate veterans from Georgia held a small reunion at Kennesaw Mountain near Atlanta. To their surprise about 320 Union veterans showed up to join them. The day featured barbecue, fireworks, and tributes to both Union and Confederate soldiers.

During the next six years 24 major reunions took place in the North and South. In 1895 over 10,000 Union and Confederate veterans gathered for a ceremonial dedication of the Chickamauga and Chattanooga National Military Park. That same year Chicago dedicated a huge memorial to honor Confederate soldiers who had been killed in the war. New York dedicated a similar memorial two years later. William McKinley, a Union veteran, was the U.S. President in 1898. Speaking in Atlanta, he said that protection of Confederate cemeteries should be a "national obligation."

The largest veterans group in the South was United Confederate Veterans (UCV). Its commander, John B. Gordon, preached full reconciliation with the United States. In 1887 he wrote:

> *[We need]* to see one more war, that we might march under the stars and stripes, shoulder to shoulder, against a common foe.

Eleven years later in 1898, Gordon's wish was fulfilled when the United States declared war against Spain. The war, fought mainly in the Pacific and in Cuba, would be known as the Spanish-American War. When that chance came many entire UCV camps volunteered to join the U.S. Army or U.S. Navy. Most of these volunteers were in their late fifties.

Military reunions became more common as death slowly thinned the ranks of surviving but aging veterans. In 1903 a Confederate memorial service in Arlington National Cemetery attracted over 2,000 veterans from the North and South. Three years later on 9 March 1906 the U.S. Congress took the first major step toward postmortem equality for Union and Confederate soldiers killed during the war. In new legislation titled, We Honor Our Fallen Ancestors, Congress authorized funds for headstones to be placed at graves of Confederate soldiers who were buried in federal military cemeteries.

Fiftieth Anniversary of the Battle of Gettysburg: Several military reunions had taken place at Gettysburg battlefield over the years. Most had been North-South gatherings of veterans from regiments or divisions. One had brought together former soldiers from the Philadelphia Brigade, of the Union Army, and Pickett's Division, of the Confederate Army. The highlight of that reunion had been a ceremony in which Gen. Lewis Armistead's sword, captured during the battle in 1863, was returned to survivors from his command.

The 50th anniversary of the pivotal struggle at Gettysburg would come in 1913. Some began to envision a grand celebration, a reunion for the ages on the old battleground. Edwin Stuart, governor of Pennsylvania, proposed an extraordinary plan to his legislature in 1909. He called for a 50th anniversary "peace jubilee" at the famous site. Two years later the U.S. Congress got involved and budgeted $150,000 for the six-day (29 June - 4 July) celebration. Congress also stipulated that the War Department (later, *Department of Defense*) would build a "Great Camp" for attendees on grounds of Gettysburg National Military Park. Thirty-three states contributed a total of $1,034,000 to supplement federal funding.

The War Department began work on the camp in early 1913. Army tents would provide housing for the 45,000 veterans expected to attend the celebration. In addition, army tents would shelter 1,400 active duty soldiers, 2,100 food service workers, 385 Boy Scouts, several thousand logisticians, and swarms of newspaper reporters. In total, army tents were provided for 57,198 persons. All honorably discharged and honorably paroled veterans of the war were invited. The commission located slightly over 40,000 such veterans and mailed elaborate personal invitations to each of them. It was thought that many more veterans probably would attend the highly publicized event. One historian noted:

> Never again in the world's history *[would]* so great a number of men, so advanced in years, *[be]* assembled under field conditions.

Near the center of the encampment a prefabricated "Great Tent" would accommodate 13,000 chairs. The U.S. Post Office set up a temporary post office. The Pennsylvania Health Department built 94 latrine and bathing facilities, and each could accommodate about 35 people at one time. The state also set up first aid stations, water fountains, and a temporary morgue. Along roads leading to the camp the American Red Cross established 14 relief and rest stations for elderly attendees.

The Great Camp opened for supper on 29 June 1913. Planners expected about 6,000 first-night attendees, but 21,000 arrived. Over the next two days additional tens-of-thousands flocked to the Great Camp via the Western Maryland Railroad, the Gettysburg-Petersburg Turnpike, and the Chambersburg-Gettysburg Turnpike. The final tally would show that 53,407 veterans attended. They came from 46 of the 48 (later, *50*) states. The reunion officially began on 1 July, a day to reminisce with old wartime comrades and establish new friendships. The next three days featured speeches, presentations, nightly fireworks displays, and camaraderie.

Each veteran who attended received a commemorative token, an engraved program, and a commemorative medal. Woodrow Wilson, the U.S. President, addressed those who could find a seat inside the Great Tent. He stressed national unity and honor for all who had suffered and sacrificed during the war. In part he stated:

> We have found one another again, as brothers and as comrades-in-arms, enemies no longer, generous friends *[instead]*, our battles long past, the quarrel forgotten.

The most anticipated event, a recreation of Gen. George Pickett's famous charge, began on the afternoon of 3 July. Fifty years after that pivotal assault in 1863, former soldiers of the North and South carried canes and crutches instead of rifled muskets. This time former Confederate soldiers, many of them old and feeble, massed for the charge only 100 feet from their objective, the famed "bloody angle." At precisely 2:00 p.m. (50 years to-the-minute after the 1863 attack had begun), they shouted at the top of their lungs and *charged.* Once more the rebel yell echoed across the battlefield as old warriors walked and hobbled forward to meet their former enemies waiting behind a stone wall. There the old adversaries no longer wielded bayonets. Instead they extended hands of friendship to their former foes. A newspaper journalist would write:

> Many of the old soldiers wept. They clasped hands and buried their faces on each other's shoulders.

Newspapers published lengthy news articles and editorials about the 50th anniversary reunion. The editors' theme was unity, and that message aroused the public conscience.

A massive granite and bronze Confederate Memorial would be dedicated the next year in Arlington National Cemetery. Woodrow Wilson, the U.S. President, delivered the keynote address. His words focused upon peace, national reunification, and sacrifices made by the common soldier.

Building a New Nation: The new country that had formed after the Revolutionary War had been a fragile and experimental union. It had been born during an era of kings and emperors. Founding Fathers had known that survival of their republic was tenuous. History had shown that most such governments had imploded internally due to revolution or anarchy. Others had fallen prey to dictators.

Before the American Civil War the United States of America had remained a loose affiliation of sovereign republics. References to the affiliation had been written or spoken with a plural verb: the United States *are* . . . . The war changed that manner of speech. The new United States had become a truly indivisible republic, and references to the nation featured a singular verb: the United States *is* . . . .

The war was a paradox. On one hand it was a destroyer of human life. On the other hand it formed the basis for most societal and technological advances in years to come. Warfare had begun near the end of the medieval medicine era, and it prompted exponential advances in surgical practices and sterility. Before the war most people had gotten medical care at home. After the war new civilian hospitals, modeled after military hospitals, cropped up in major cities.

Begun in 1863, the transcontinental railroad had been intended to bind California to the Union during the war. The war spawned the land-grant college system. Born as a result of the war, Memorial Day (originally, *Decoration Day*) evolved into a national observance and holiday. The war outcome finally settled the constitutional secession question, and after the war the Thirteenth Amendment ended chattel slavery nationwide.

Before the war the states had shared a federal government that had almost no contact with the average citizen. Exclusive of federal post offices, most citizens had never dealt with the U.S. Government. However, after the war the United States began taxing citizens directly. National reliance upon state militias became secondary, for the revitalized nation established federal armed forces. A new federal banking system, new national currency, and increased power of federal courts became hallmarks of post-war America.

> Flames of the Civil War forged the framework of modern America.
> *[James McPherson, Out of War, a New Nation, 2010]*

After the war ended in 1865 the remaining 35 years of the century had witnessed astounding technological and social change. Marriage of technology and industry accelerated. The 1870s had witnessed the first practical telephone and first commercially viable electric light bulb. High-strength steel enabled construction of the first skyscraper, a ten-story building in Chicago. Electrically powered elevators soon paved the way for even higher buildings.

The first practical typewriter, modern cash registers, the motion picture camera, electric voting machines, and the first transatlantic telegraph cable debuted. Levi Strauss developed and marketed blue jeans with brass rivets. John Pemberton developed the secret Coca-Cola formula. Marconi Guglielmo invented a *wireless*, later to be called a radio. New medical technology embraced x-rays. Gasoline-powered farm tractors began replacing horses and mules.

Use of rabies and anthrax vaccines became common. The famed Brooklyn Bridge had opened for traffic in 1883. Basketball, a new game, became wildly popular at the college level. The first Ferris Wheel thrilled thousands, and United States Steel Corporation (aka, *U.S. Steel*) and Standard Oil forged industrial empires.

The first "motor car" powered by a gasoline engine had been sold in 1886, and within two decades Ford Motor Company would be mass producing the Model A on an assembly line. By the turn of the century in 1900, Pennsylvania was producing half of the entire world's oil supply. Factory jobs began to replace farming and home crafts. An industrial economy based upon free enterprise and free labor became the national norm. Factory work and clerical office jobs replaced farming as the most common occupation for men.

Veterans Lead the Nation: Veterans played the crucial role and led the nation in a reconciliation process. Surviving veterans had seen friends and comrades die in battle. After the war their visits to military cemeteries became humbling emotional experiences. In honor of the dead most veterans resolved to do their best to make their nation a better place. Their resolutions would be exemplified in words of a eulogy delivered by an American military chaplain in March 1945. He would speak at a new American military cemetery on a tiny island in the vast Pacific Ocean:

> Here lie bodies of comrades and friends. . . . Among these men there is no prejudice, no hatred. Theirs is the highest and purest democracy.
>
> Thus do we memorialize those who, having ceased living with us, now live *within* us. Thus do we consecrate ourselves, the living, to carry on the struggle they began.
>
> We here solemnly swear! This shall not be in vain! Out of this *[war, and]* from the suffering and sorrow of those who mourn will come, we promise, the birth of a new freedom for the sons of men everywhere. Amen.

Surviving veterans of the American Civil War provided political leadership, promoted national reconciliation, and set examples of hard work and perseverance. Six of them were elected to the office of U.S. President. Others became governors and college presidents. Still others became state and local civic leaders. They became prime examples of selfless dedication to their family, their community, their state, and their nation.

Veterans of the brutal war, from the North and from the South, had been shaped by their common experience of shared hardship and commitment to cause. After the war they got jobs, worked, started businesses, voted, raised families, paid taxes, and together they created a unified society. They forged the national framework that led the United States into the Twentieth Century as a world power.

Recognition, Honors, and Unity: For a half-century after the tragic Reconstruction Era, nowhere had national unity been more in evidence than among former Confederate leaders. A bronze statue of Jefferson Davis, the former Confederate president, was erected in the U.S. Capitol building in Washington, DC. Nearby stands a statue of the former Confederate vice-president, Alexander Stephens. Also in the capitol building stand statues of three former Confederate Army generals: Robert E. Lee, Wade Hampton, and Joseph "Fighting Joe" Wheeler. After the long Reconstruction Era, former Confederate military and political leaders served in both houses of Congress. Fifteen former Confederate Army officers served as United States ambassadors and as ministers in foreign countries.

Wheels of government grind slowly, but on 26 February 1929 the U.S. Congress expanded military burial honors policy to include all Confederate veterans. The new law authorized the Secretary of War (later, *Secretary of Defense*) "to erect headstones over the graves of soldiers who served in the Confederate Army" regardless of their place of burial (prior legislation in 1906 had limited Confederate headstone placement to graves in federal military cemeteries). This new law made Confederate military headstone benefits equal to those afforded to Union military veterans.

On 23 May 1958 the U.S. Congress approved a symbolic gesture. Each person who served in the Confederate Army or Navy was granted a federal monthly pension "in the same amount and subject to the same conditions" as an equivalent United States soldier or sailor. However, this new legislation was designed only as a symbol of military equality. It had no financial impact, for the last verified Confederate veteran already had died.

Congress wished to honor and recognize Confederate soldiers as the one-hundredth anniversary of the war drew near. The lawmakers prodded the U.S. President, Dwight D. Eisenhower, to formalize such honors and recognition. On 6 December 1960, 100 years after the first southern state had seceded, the congressional wish was publicly fulfilled. The presidential proclamation states in part:

> The *[American Civil War]* was America's most tragic experience. But like most truly great tragedies, it carried with it an enduring lesson and a profound inspiration. It was a demonstration of heroism and sacrifice by men and women of both sides who valued principle above life itself, and whose devotion to duty is part of our Nation's noblest tradition.
>
> Both sections of our now reunited country sent into their armies men who became soldiers as good as any who ever fought under any flag. . . . That America could produce men so valiant and so enduring is a matter for deep and abiding pride.
>
> . . . *[The war was]* a lasting testimonial to something unconquerable in the American spirit. A transcending sense of unity and larger common purpose, in the end, caused men and women who had suffered so greatly to close ranks, once the contest ended, and to go on together to build a greater, freer, and happier America and must be a source of inspiration as long as our country may last.

For over 50 years after the American Civil War, former Union and Confederate soldiers – common men – formed the backbone of their nation. Working together, they left an indelible constructive mark on society. By virtue of the sacrificial wartime and post-war lives they lived, their place in history became honored and respected. No other group of citizens has made a greater contribution to the United States of America.

> We are all lucky, because we live in the best-est country in the whole world. God bless America!
> *[a Third Grade girl, speaking during a Veterans Day program at her school near McCormick, South Carolina, 11 November 2008]*

Union and Confederate veterans shake hands at the 50th anniversary "Peace Jubilee" reunion at Gettysburg battlefield in July 1913.

A portion of the Great Camp at Gettysburg battlefield for the 50th anniversary "Peace Jubilee" reunion in July 1913.

A Confederate veteran and Union veteran shake hands at the 50th anniversary "Peace Jubilee" reunion at Gettysburg battlefield.

Confederate Memorial, Arlington National Cemetery.

– finis –

– *About the Author* –

The author served as a combat-assault helicopter pilot in the U.S. Marine Corps during the war in Vietnam. Thereafter he became a commercial pilot for the oil and gas industry. Later he worked as a manager at AT&T for 25 years, and during ten years of that period he served as a guest instructor at Bell Communications Research.

After retirement the author began researching military history, aviation history, and regional history. A collection of his personal correspondence is held by the Manuscript Division at the South Caroliniana Library *(Marion F. Sturkey Papers, 1963-2006)*. The author now lives in his original hometown, Plum Branch, South Carolina. This is his 15th book.

– *Index* –

Note: In the interest of clarity this index is presented in two parts:

Part One consists of publications such as books, magazines, newspapers, broadcasts, etc.

Part Two is a listing of names of persons, events, and institutions.

– Part One –

A Rumor of War (book). 239
American Rifleman (magazine). 213, 216, 219
Augusta Chronicle (newspaper).. . . . . 217-218, 221
Battles and Leaders, Vol. II (book). 98-99, 106
Berry Benson's Civil War Book: Memoirs of a Confederate Scout and Sharpshooter (book). 211-221
Boston Journal (newspaper). 99
By Force of Arms (book). 138
Century (magazine). 217
Charleston Mercury (newspaper). 130
Columbus Dispatch (newspaper). 251
Confederate Veteran (magazine). 86
Constitution of the Confederate States (document). 152, 154
Constitution of the United States (document). 152, 157
Cotton: From Farm to Fabrics (treatise).. 4
Diary of a Southern Refugee during the War (book).. 109-113
Dixie (song). 151
Edgefield Advertiser (newspaper). 227, 235
Farmers' Register (magazine). 150
Harpers Weekly (magazine).. . . . . 137, 240
Holy Bible (holy book).. 151
How Few Remain (book). 138
Lee's Lieutenants (book). 99
Lexington Gazette (newspaper). 136
Liverpool Mercury (newspaper).. 179
Maxims of War (book). 132

Military Review (magazine). 128
Myths About Slavery (treatise). 158
New Georgia Encyclopedia (reference). 220
New York Tribune (newspaper). 146
New York World (newspaper). 235
Notes on the Next War (book). 208
Ordinance of Nullification (document). 7-8
Ordinance of Secession (document). 15
Out of War, a New Nation (treatise). 246, 264
Outlook (magazine). 217-218
Recessional (poem). . . . . viii, 199
Richmond Dispatch (newspaper). 22
Richmond Examiner (newspaper). 130
Richmond Whig (newspaper). 51
Seal of the Confederate States (legal imprint). 29, 36
South Carolina, a History (book). 224
South Carolina Exposition and Protest (document). 7
Southern Historical Society Papers (collection). 87
Stonewall Jackson at Gettysburg (book). 138
Tariff of 1928 (document). 6
The Art of War (book). 127, 138
The Birth of a Nation (motion picture). 241
The Charge of the Light Brigade (poem). 247
The Civil War (television series). 221
The Prescript (secretive oath). 223
The Story of Stonewall Jackson (book). 141
These Good Men (book). 259-260
Trans-Atlantic Slave Trade Database (data collection). 148
Uncle Tom's Cabin (book). 151
Washington Post (newspaper). 133

– Part Two –

– A –

Achilles. 63
Adair, Lyle. 79
Alexander, Edward. 130
Amnesty Oath. 32-33, 85
Anaconda Plan. 24

Andersonville POW Camp. 71-72, 77-82, 91
Antietam, Battle of (aka, *Sharpsburg*). 97-98
Armistead, Lewis. 261
– B –
Baker, Mark. 248
Balaklava, Battle of. 247
Banks, Nathaniel. 126
Barry, John.. 133
Beall, Lloyd J.. 173
Bee, Barnard.. 21
Bell, H. Lee. 80
Belle Isle POW Camp. 76, 80
Benjamin, Judah P.. 33
Benson, Abraham.. 211
Benson, Berry.. . . . . 23, 84-85, 211-223
Benson, Blackwell. 212
Benson, Nancy. 211
Benson, Susan Williams. 221
Bilby, Joseph G.. . . . . 213, 216, 219
Bixby, Lydia.. 119
Blood Stained Banner (flag). 28
Bly, Douglas.. 249
Bolden, Vernie.. 141
Bonaparte, Napoleon. 42, 131-132, 136
Bonnie Blue Flag. 27-28, 38
Booth, John Wilkes. 207
Boulanger, Gustave. 159
Bradley, Omar. 128
Bragg, Braxton.. 33
Brake, Polly. 122
Breckinridge, John C.. 33
Breckinridge, W.C.. 33
Briggs, Eliza.. 3
Briggs, Henry.. . . . . 1, 246
Briggs, John.. 1
Briggs, Pinkney. 22
Briggs, Richard P.. . . . . 2, 246
Briggs, Thomas J.. . . . . 2, 246

Brown, Henry. 157-158
Brown, John. 125
Buchanan, Franklin. 185
Buchanan, James. 18
Bull Run, First Battle of (aka, *Manassas*) . . . . . 21-23, 43, 76, 103, (continued) 113, 126, 247
Bull Run, Second Battle of (aka, *Manassas*). 128, 212
Bulloch, James. 168, 170, 181-182
– C –
Calhoun, John C.. . . . . 6, 10, 12
Campbell, Randolph. 226
Caputo, Philip. 239
Cashin, Edward J.. 211
Chadwick, Bruce. 13
Chamberlain, Daniel. 237-238
Chamberlain, Joshua L.. 243
Chancellorsville, Battle of. 131-134, 212-213
Chase, William.. 141
Chickamauga, Battle of. 247, 260
Chimborazo General Hospital. 52-53
Chittendon, Lucius.. 76
Clark, M.H.. 34
Cobb, Thomas. 218
Coker, Rachel.. 245
Cold Harbor, Battle of. 40
Columbus, Christopher. 148
Cotton Kingdom.. 4-13
Courtenay, Thomas. 206
Crawford, James. 215
Cromwell, Oliver. 164
Curtis, William.. 55
– D –
Davis, Jefferson. 6, 17, 29, 33-34, 44, 140, 165, 167, 178, 200, 266
Davis, Thaddeus.. 84
de Turenne, Vicomte. 200
Decatur, Daniel. 151
Democratic Club of Liberty Hill. 231
Dibrell, George C.. 33

Dinnerstein, Leonard. 220
Dix-Hill cartel. 70
Dorn's Mine Democratic Club.. 230
Douglas POW Camp. 75, 77
Downing, Lylburn. 140-141
Drake, Francis. 194
Duke, Basil W... 33
– **E** –
Elmira POW Camp. 71, 77, 82-88, 90, 215, 219-220
Evans, Henry. 86
Eagle (observation balloon).. 132
Ellison, William.. 151
Edgar, Walter.. . . . . 224, 236, 238
Eisenhower, Dwight D.. 266
– **F** –
Fall, Bernard B.. 43
Farragut, David. 172
Ferguson, S.W... 33
Fhagan, Mary. 219
Flavion, Gary. 77
Fontana, Antonio. 218
Foote, Shelby. 219
Forrest, French. 184
Forrest, Nathan Bedford.. . . . . 233, 258
Fort Fisher. 84, 172
Fort Sumter. 19, 36
Frank, Leo. 219-220
Frazier, Garrison. 155
Fredericksburg, Battle of. 113-114, 130, 212
Freeman, Thaddeus. 22
Fremont, John. 154
– **G** –
General Hospital No. 1. 56
General Hospital No. 2. 56-57
General Hospital No. 3. 57
General Hospital No. 4. 57
General Hospital No. 5. 57
General Hospital No. 6. 57

General Hospital No. 7. 57-58
General Hospital No. 8. 58
General Hospital No. 9. 58
General Hospital No. 10. 58-59
General Hospital No. 11. 59
General Hospital No. 12. 59
General Hospital No. 13. 59-60
General Hospital No. 14. 60
General Hospital No. 15. 60
General Hospital No. 16. 60
General Hospital No. 17. 60
General Hospital No. 18. 61
General Hospital No. 19. 61
General Hospital No. 20. 61
General Hospital No. 21. 61-62
General Hospital No. 22. 62
General Hospital No. 23. 62
General Hospital No. 24. 62
General Hospital No. 25. 62
General Hospital No. 26. 62-63
General Hospital No. 27. 63
General Hospital No. 28. 63
George V, King. 252
Gettysburg, Battle of. . . . . . 30, 103, 113, 118, 138, 213, 247, 256, (continued) 261-263, 268-269
Gibbony, Douglas. 139
Gordon, John B.. 260
Gorgas, Josiah. 170
Grant, Ulysses. 70, 153, 233, 235, 237
Grayson, John.. 168
Greeley, Horace. 146
Gress, Samuel. 45
Guglielmo, Marconi. 264

– H –

Hacker, David. 245
Hamburg Riot.. . . . . 218, 236
Hammond, James.. 4
Hampton, Wade. 22, 237, 266

Harmon, John L.. 235
Hatch, Reuben. 201, 206
Hayes, Rutherford. 237
Hemmingway, Ernest.. 208
Hill, Daniel.. 128
Hoffman, William. 82-84
Holt, Daniel. 45
Homer. 63
Hooker, Joseph.. 131-132
Hopkins, Maudie. 254
Horwitz, Tony. 138
Houston, Ulysses. 156
Howard's Grove General Hospital.. 53-54
Hunley, Horace L.. 179-180
Hunter, Alexander. 50
Hunter, David.. 154

– J –

Jackson, Anna. 124, 129, 135
Jackson, Cummins. 122
Jackson General Hospital.. 54-55
Jackson, Gilmer. 215
Jackson, Julia. 121-122, 124, 129
Jackson, Mary. 124
Jackson, Thomas "Stonewall". . . 21, 26, 48, 54, 121-141, 212, 218
James, George S.. 19
Jefferson, Thomas. 230
Jemison, Edwin F.. 39
Jennings, Wesley. 22
Johnson, Andrew. 156
Johnson, Anthony. 149
Johnson, FNU. 22
Johnson, Samuel. xiii
Johnson's Inland POW Camp. 75
Johnston, Joseph E.. . . . . 178, 207, 217
Junkin, Elinor.. 124

– K –

Kellog, Robert H... . . . . 71, 78
Kesler, William. 86-87

King, John A.. 87
Kipling, Rudyard. . . . . . viii, 199
– L –
Lacy, Tucker. 135
Lee, Fitzhugh. 132
Lee, Robert E.. . . . . 26, 32, 127-131, 135, 140, 178, 217-218, 225, (continued) 258, 266
Lewis, R.A.. 55
Libby POW Camp. 75
Lincoln, Abraham. . . . 9-10, 17, 19, 32, 69-70, 119, 134, 138, 146, (continued) 153-155, 158, 207
Logan, John. 251
Longstreet, James.. . . . . . 128, 130
Louden, Robert.. 205-206
– M –
Maffitt, John.. 170
Mallory, Stephen R.. . . . . . 33, 167-168
Malone, S.C.. 215
Malvern Hill, Battle of. 39
Marschall, Nicola.. 27
Marvel, William.. 79
Mason, Cass.. 200-207
Maull, John. 215
Maury, Matthew.. 228
Maximilian (emperor and dictator). 32, 228
McCarthy, Harry. 27
McCaw, James B.. 53
McClellan, George.. 129
McDuffie, George. 6
McGuire, Hunter. 48, 133-135
McGuire, Judith W.. . . . . . 93, 109-113
McKim, Randolph. 1
McKinley, William. 260
McNeill, Alex. 30
McPherson, James. 244, 246, 264
Medical College of Virginia. 49
Meredith, S.. 55
Miltimore, John. 10

Minie, Claude-Etienne. 42
Morrison, Anna. 124
Mosby, John. 41
Moss, Franklin J.. 236
Moxley, Tonia. 141
– N –
Newton, Nathan. 110
Newton, Thomas D.. 117-118
Norman, Michael. 259
Norris, Albert. 208
Norris, William. 229
North Anna, Battle of. 56
– O –
O'Hara, Theodore. 35
Old Capitol POW Camp. 81, 215
– P –
Page, Thomas J.. 175-176
Parker, Niles G.. 236
Payne, John. 180
Pemberton, John. 264
Pickett, George. 262
Point Lookout POW Camp. 76-77, 214-215
Ponzi, Carlo. 220
Pope, John. 128
Porter, David. 172
Pruitt, Sarah. 158
Putegnat, John. 215
– R –
Reagan, John H.. 33
Rhea, Gordon. 151
Rochelle, M... 22
Ruane, Michael E.. 133
Rugg, George. 81
– S –
Salisbury POW Camp. 75-76
Salzbach, Battle of. 200
Scott, Robert K.. 236
Scruggs, J.P... 215

Sears, Stephen W.. 130
Secession Convention. 14
Semmes, Raphael.. 181-184
Seven Days Battles.. . . . . 127, 212
Seward, William. 32
Sharpsburg, Battle of (aka, *Antietam*). . 94, 128-129, 212, 247, 255
Shelton, Glenn. 215
Sherman, William T.. . . . . 88, 155, 176
Shiloh, Battle of. 247, 251
Slaton, John. 219
Slidell, John. 174
Smith, Kirby.. 178
Speed, Frederick.. 207
Stainless Banner (flag). 28
Stansbury, FNU. 111
Stanton, Edward.. . . . . 155, 186
Stars and Bars (flag).. . . . . 27-28, 38
Stephens, Alexander. 32, 266
Stone, Jesse. 22
Stratton, Alfred A.. . . . . i, 64
Strauss, Levi.. 264
Stuart, Edwin. 261
Stuart General Hospital. 55
Stuart, J.E.B. “Jeb”. 21, 55, 129
Sturkey, William O.. 86
– **T** –
Tansill, Robert. 173
Templin, William.. 215
Tennyson, Alfred. 247
Tilden, Samuel.. 237
Traweek, Washington. 85, 215
Tzu, Sun.. . . . . 127, 130
– **V** –
Vaughn, J.C... 33
– **W** –
Waddell, James. 177-178
Walker, William.. 218
Wallen, Milton.. 66

Washington, George.. . . . . 29, 36
Wayne, John (aka, *Marion Morrison*). 242
Wheeler, Joseph. 266
White, William.. 125
Whitman, Walt.. . . . . 42, 46
Whitney, Eli. 4, 11, 149
Williams, George. 201
Williams, Susan. 221
Williams, Walter E.. 147
Wilson, Woodrow. 262-263
Winder General Hospital. 55-56
Winder, John. 55
Wirz, Henry. 72, 80-81
Wood, Fernando.. 137
Woolson, Albert.. 254
Worden, John.. 186

– Current Books by Marion Sturkey –

Warrior Culture of the U.S. Marines (3rd Edition)

No Greater Love: The Groucho Marx Battle

Bonnie-Sue: A Marine Corps Helicopter Squadron in Vietnam (2nd Edition)

Murphy's Laws of Combat (2nd Edition)

Mayday: Accident Reports and Voice Transcripts from Airline Crash Investigations

Mid-Air: Accident Reports and Voice Transcripts from Military and Airline Mid-Air Collisions

Military Monuments and Memorials in South Carolina

Plum Branch: Heaven in South Carolina

Gone But Not Forgotten (2nd Edition)

Casualties of War

– Out of Print –

Gone But Not Forgotten (1st Edition)

Gone But Not Forgotten: An Introduction

Bonnie-Sue: A Marine Corps Helicopter Squadron in Vietnam (1st Edition)

Murphy's Laws of Combat (1st Edition)

Warrior Culture of the U.S. Marines (1st and 2nd Editions)